GREAT
BALLS OF FIRE
A Unified Theory

of ball lightning, UFOs, Tunguska and other anomalous
lights

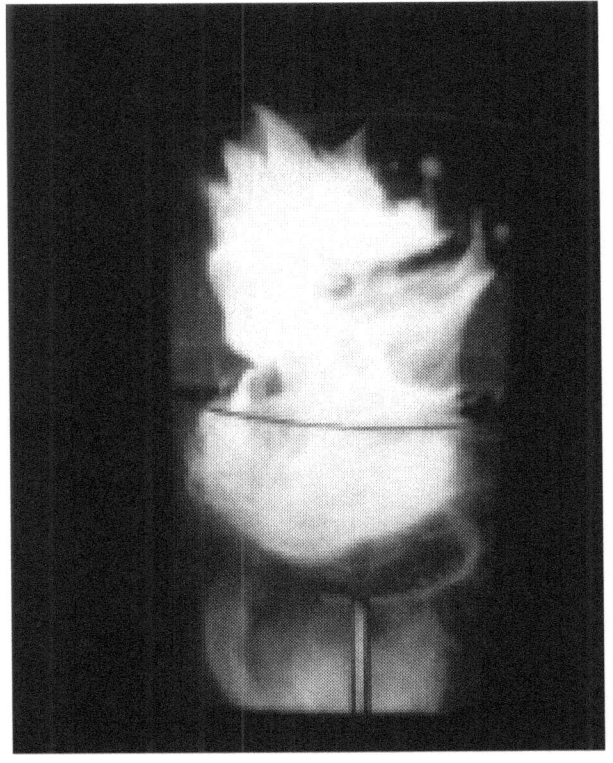

Peter Francis Coleman MSc, MAppSci

Published by
FIRESHINE PRESS,
Christchurch
New Zealand

This is a revised second edition to that previously published in 1997
under the title
"Ball Lightning- a scientific mystery explained"
ISBN 0-473-04827-2
by Fireshine Press,
Christchurch
New Zealand

CONTENTS

*To my loving wife
Sue,
and to my children,
Joe, Luke,
and
Annabelle*

PREFACE

"Forth rush'd from whirlwind sound
The chariot of Paternal Deity,
Flashing thick flames, wheel within wheel, undrawn..."
Milton

This book contains what I believe to be the correct scientific answer to the problem concerning fireballs, or "chariots of fire" that have appeared in our skies over past millennia. Several authors have made this claim but not one of their theories advanced is as capable of explaining the eyewitness observations as the vortex fireball theory.

The dual puzzle of ball lightning and a type of UFO has managed to endure for at least two centuries despite the rapid development of sophisticated scientific tools both mathematical and technical. The fireball phenomenon itself often manifests itself as a luminous sphere appearing in the Earth's lower atmosphere. These fireballs are certainly not meteors, though some observers continue to perpetuate this explanation. Anecdotal descriptions of this fireball are large in number and are scattered throughout the literature. Although several some prominent scientists have tried to unravel this conundrum, not a single hypothesis has been widely successful from the many currently available. Arago, a prominent scientist in 1855 commented on the difficulty of the problem;

" One of the most inexplicable problems within the range of physics."

Some scientists (this includes my supervisor for my Master's research) have theorized about ball lightning's nature without having realized the bigger picture involved, one directly connected with hard-to-explain UFOs associated with the extraterrestrial hypothesis. There are many physics theories that are spin offs from their own fields and do not address the field observations collected from the literature.

The idea that UFOs could be explained with ball lightning is not new-the notion has been around for sometime. Carl Benedick's work " *Theory of lightning balls and its application to the atmospheric phenomenon called "flying saucers"* was published in 1951 at the height of the saucer scare. Better known are the protagonists in UFO history, such as Klass, whose theory of ball lightning used the commonly held idea that ball lightning was electrical, hence his corona theory of UFOs. But such theories were dismissed by UFO researchers. The theories were simply incapable of explaining the diverse observations reported. If these theorists

5

had the correct theory as to the nature of ball lightning they might have been able to defend their theories and convince others as well.

I am indebted to the work of other researchers who have provided the source material to piece together the jigsaw. There are so many detailed aspects to this theory that could be fruitfully pursued by workers in the field. What I have done here is to advance the theory as a generalized first step. The research is an ongoing process and every now and again a new piece of supporting evidence or prediction comes to light. While I was writing a paper (Coleman, 2005) for the Journal of Scientific Exploration I came across new references that helped the case for the theory. Reports of a series of ball lightning seen to separate from a tornado spout as though the funnel was a fireball manufacturing plant do exist. According to my theory these fireballs are smaller vortices that have 'spun' off from the main funnel. I would therefore expect non-combusting vortices to 'hatch' from the funnel bottom. Sure enough I found the evidence. A meteorological observer (Justice, 1930) had indeed seen smaller vortices that detached from the funnel tip.

The positive statements about the validity of the theory from scientists are encouraging and it is just a matter of time for full acceptance. My theory has also been cited by a number of professionals in the field. In one case a Russian scientist, Vladimir Svetsov from the Russian Academy of Sciences, who I corresponded with, used my theory to account for a peculiar 'meteor' called the 'Santiago Compostela' event in a chapter of a report to the Sandia Laboratories in the US.

By publishing this book my theory is now open to all readers to see for themselves whether the solution matches what ball lightning and UFO observers have been seeing all along.

PART I

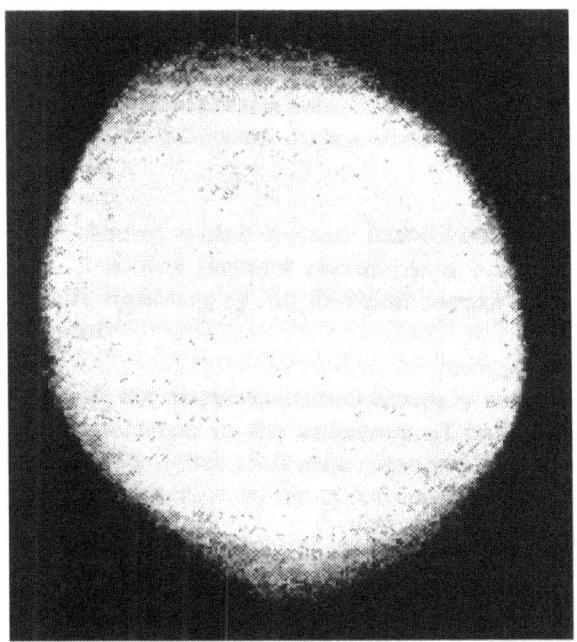

BALL LIGHTNING

A ball lightning photograph widely considered to be authentic
Source: Norinder (cited by Singer, 1971)

1
The Enigma of Ball Lightning

GLOBE OF FIRE DESCENDING INTO A ROOM.

Ball Lightning - "Globe of Fire Descending into a Room" in "The Aerial World," by Dr. G. Hartwig, London, 1886. P. 267. [Courtesy NOOA Photograph library]

"It is one of the pleasures of meteorology that many of its phenomena are still imperfectly understood. Few, however remain such a complete enigma as ball lightning BL, or Kugelblitz, the name given to the mobile, long-lived, luminous spheres which, for centuries, have been occasionally reported in the vicinity of thunderstorms and elsewhere."
Charman (1982)

BALL LIGHTNING'S NATURE

Ball lightning is a longstanding problem in science and although many common phenomena have universally accepted explanations, ball lightning has no solution that has been widely accepted. Stars shine due to light emission from nuclear reactions and Newton discovered the nature of the rainbow and something about the nature of light itself. By experiment he showed that white light is made up of several different colors or wavelengths of light. The fact that ball lightning has proved to be such an intractable scientific problem for so long is intriguing. Since no one has known what it is then no one can predict with certainty when and where it will take place.

There are many theories: ball lightning could literally be anything from black holes to self-contained plasma. The current situation compared to that of over a hundred years ago is therefore no different. One frustrated ball lightning investigator, the French scientist, Camille Flammarion, elegantly enunciated the situation in 1905 regarding the problem of ball lightning. He wrote:

"But we must confess that if spheroidal lightning seems particularly capricious, it is because we are still ignorant of the laws which guide it. Our ignorance alone is the cause of the mystery. We try to discover the enigma in the silence of the laboratories, where physicists question science without ceasing; we try to reproduce the fireballs artificially, but the problem is complicated, and its solution presents enormous difficulties."

Much later a physicist and popular science writer, Professor Paul Davies (1987), summarized the state of ball lightning research in a 1987 December issue of *New Scientist*, as follows:

"The phenomenon responsible for these manifestations-ball lightning- is one of the more bizarre scientific riddles of our age. The idea that balls of fire can pop out of nowhere, meander around for a while putting the fear of God into people, and then disappear again, seems patently absurd. As recently as the 1970s, some physicists maintained that lightning balls were nothing more than spots before the eyes. But the fearsome things have stubbornly refused to go away. Furthermore, we cannot attribute all reports to deluded laymen. Eyewitnesses include no lesser scientists than Niels Bohr, Victor Weisskopf and Martin Ryle."

Early ball lightning scientific research.

The formal discussion of the problem of ball lightning is said to have begun with Arago in 1837, although Lord Kelvin doubted its existence, and Faraday thought that if the phenomenon did exist, it was not electrical. Arago brought ball lightning to the attention of the French Academy of Sciences. Arago (1838) in his *"Sur le Tonnerre"*, wrote,

"... globular lightning, which appears like a luminous ball or globe of fire; it moves through the air at a comparatively slow rate, while lightning of the first and second class exist but for a moment."

Although Arago seems to be saying that ball lightning exists but Tomlinson's (1889) letter to *Nature* suggested he is somewhat skeptical.

"Another point of interest in this valuable communication is the introduction of ball lightning. Arago is skeptical as to the existence of ball lightning (eclairs en boule), or that which moves through the air at a comparatively slow rate, appearing like a luminous ball or globe of fire. Faraday is equally skeptical. But the well attested cases of what we name ball lightning, and the Germans "Kugelblitz", are so numerous that they can no longer be termed in Arago's language, "a stumbling block (pierre d'achoppement) for meteorologists."

This 1889 quotation is possibly one of the earliest uses of the term "ball lightning". It clearly stated that the term is English. "Globular lightning", semantically similar to the term "ball lightning", was also used by the English (Hare, 1889).

The term "ball lightning" continues to be widely used in the English speaking world, but it is an unfortunate misnomer because the very name implies that the phenomenon is somehow directly connected with lightning. But an unknown phenomenon does not logically need to be connected with lightning. In fact contemporary research shows that ball lightning is also seen in fine weather (without fair-weather lightning) contradicting the exclusive association with lightning. Some investigators have gone even a further step and defined ball lightning as an electrical phenomenon-far from it. Both assumptions are unsupported by observation.

It is interesting to reflect on the extraordinary progress of lightning research with the complete lack of progress of ball lightning. It was Benjamin Franklin (1753) who correctly deduced the true nature of ordinary lightning. He showed that lightning was no different in substance to electrical discharges that he had observed in his experiments. The nature of ball lightning has remained an unsolved problem from around the time science was in its infancy.

The reported method by which Franklin demonstrated the nature of lightning was to bring lightning down from the sky, via a tethered kite. The lightning followed a conduction path down along string to a metal key. The metal key acted as a conductor to transfer electric charge from lightning which then charged up a "phial" or "Leyden jar", which is a type of capacitor to store electrical charge. What Franklin found was that the same type of experiments he usually carried out with static electricity could also be carried out with the lightning charge stored in the Leyden jar. Thus, the nature of lightning was demonstrated as another case of what he termed "electric fire". Franklin (1753) wrote:

"At this key the phial may be charged, and from the electric fire thus obtained, spirits may be kindled, and all the other electrical experiments may be performed, which are usually done by the help of a rubbed glass rod or tube, and thereby the sameness of the electric matter with that of lightning completely demonstrated."

By all accounts Arago was the first scientist to consciously bring ball lightning into the province of science, as a problem worthy of rational inquiry (Singer, 1971). Powell and Finkelstein (1970) claimed that Professor Georg-Wilhelm Richman, (died in 1753), was the first scientist to study ball lightning. But this needs clarification. Although he did see a pale blue fireball after lightning struck a metal rod, the ball killed him. This was while he was investigating ordinary lightning. He was repeating Bejamen's Franklin's sentry box experiment on lightning. Richman could hardly be said to be purposefully "studying" ball lightning itself. Though Richman may not have been the first scientist to see ball lightning, or to be purposely investigating ball lightning directly, he entered the history of ball lightning.

Some scientists have gone on record as being skeptical as to the actual existence of ball lightning. But this school of thought is diminishing. Hill (1960), when at the Lightning and Transients Research Institute in the United States, re-emphasized the reality of ball lightning's existence:

"Their peculiar physical behavior, coupled with the difficulty that has been encountered in attempts to simulate them under controlled conditions, has caused many scientists to adopt a skeptical attitude toward their existence. Nevertheless, the number of well-attested cases now known seems to be large enough to warrant their consideration as genuine physical events having a certain measure of reproducibility."

Fortunately the situation regarding the lack of reliable ball lightning observations has improved. There are now larger surveys and databases from which to draw upon. The general consensus among many contemporary scientists is not whether ball lightning exists. The next step is to uncover the true nature of ball lightning. Unfortunately a popular view, widely disseminated in many reference books, like encyclopedias, is that ball lightning is a peculiar and exotic form of lightning. As I have said earlier, this is an unsubstantiated assumption.

Many investigators have attempted to solve the riddle of ball lightning's nature but there has been no real progress since Arago's formulation of the problem. In a short article Dr Stanley Singer summarized in *Nature* the state-of-the-art ball lightning research (Singer, 1963).

"The problem of ball lightning has received much attention, including

numerous experimental investigations; yet the available information is even more limited (than ordinary lightning). The experiments have failed to duplicate major properties of the natural phenomenon, and it can only be said that promising explanations have been proposed."

The existing view among many scientists, engineers and physicists, is that ball lightning is a major scientific enigma. It is my hope that this book may help to bring the vortex fire ball theory to center stage in the scientific world. In the following chapter I will examine some of the competing theories to account for ball lightning.

2
Ball Lightning Theories

*Ranada and Trueba (Ranada, A.F., Trueba, J.L., 1996, Nature, **383**, 32) in a much-publicized paper suggested that ball lightning is formed from knots in the electric or magnetic components of an electromagnetic field. From this they derived a theoretical model which bears no resemblance to ball lightning (**Jennison, 1997**).*

A plethora of theories

Theories on ball lightning are being published every year. Barry and Singer (1988) stated that new theories of ball lightning have averaged five papers per year in a period from 1971 to 1980. At a guess, the number of different ball lightning theories that have been published may number in the thousands. I have a bibliographical guide of ball lightning, for the period 1982 to 1992, compiled by A.I. Grigoryev and T.N. Dunaeva from Yaraslavl University, Sovetskaya in Russia. They catalogued 693 articles published on ball lightning. The types of ball lightning theory, currently available, are wide-ranging and mostly unfamiliar to the non-professional. It is evident many imaginative and esoteric schemes have been proposed for ball lightning, irrespective of whether the theory is in agreement with the actual evidence or not. The "Atmospheric Rydberg Matter" theory and the "polymer net" theories are examples in this category but there are many more.

IS BALL LIGHTNING AN OPTICAL ILLUSION?

The theory of after-images

There are still a few dissenters who doubt the material existence of ball lightning but that their number are diminishing Some hypotheses treat ball lightning as merely an optical illusion. One theory says that ball lightning is merely spots before the eyes caused by after-images left on the retina from a bright light source, such as lightning (Argyle, 1971). This theory was discussed in a series of letters to *Nature* in 1971 (Jennison, 1971; Charman, 1971; Davies, 1971). All three authors are prominent scientists and took exception to this physiological explanation. One reason in favor of ball lightning's existence is the personal experience of one correspondent who actually witnessed ball lightning first hand. Professor Jennison of the Electronics Laboratory at Kent University observed a ball lightning object bobbing down an aircraft cabin aisle. He wrote:

"No one will deny the existence of afterimages and most people have experienced them on several occasions, but afterimages which clearly emerge from doorways, are occulted by nearby objects, yet obscure those behind; which have angular diameters increasing from five to six degrees to over twenty degrees as they pass from over two meters to within fifty centimeters of the observer, require remarkable properties of the retina or extraordinary hallucinations on the part of the observer."

A letter to Nature (the same issue) by an English optical expert, W.N. Charman from Manchester University listed serious problems with the after-image thesis. He pointed out that this optical illusion theory certainly could not explain the reports of odors or color changes. Furthermore, real ball lightning sightings generally do not remain constant in color as would be expected from images retained on the retina.

CLASSIFYING BALL LIGHTNING THEORIES.

External energy input or self-contained theories

Ball lightning hypotheses are broadly categorized into two types. The first type of theory assumes that the energy content of the fireball is self-contained while the second kind claims that energy of the fireball comes from the external surroundings. Self-containment theories propose that all the energy of the ball is introduced at the time of creation. Self-containment theorists are forced into proposing unrealistically high energy densities if they consider classic cases like the ball lightning the size of an orange that fell in a barrel of water and boiled it for several minutes (Goodlet, 1937). The assumption of these theories is that all the energy is contained in the ball at the time of creation without it being supplied from the outside. This is a real difficulty because very high energy densities are required in some cases. On the other hand, theories where energy is continually fed from the outside to the ball have a distinct advantage in overcoming the apparent high energy density problem.

Singer's classification.

Singer carried out the task of classifying many ball lightning theories in his 1971 book, *The Nature of Ball Lightning*. He classified ball lightning theories into twelve categories (Singer, 1971). The twelve categories are: agglomeration theories, Leyden jar structures, transformation of linear lightning into ball lightning, chemical theories, nuclear theories, charged dust and droplet theories, molecular ion clouds theories, vortex structures theories, electrical discharge theories, luminous spheres from vaporized substances, plasma theories, and natural electromagnetic radiation theories. His scheme covers most of the theories ever proposed to account for ball lightning.

Singer's book examined each of the twelve classes of theories and outlined the types of difficulty connected with each class of theory as well as some of the advantages. To make progress with any particular class of theory, these problems, whether real or perceived, need to be discussed and overcome.

An alternative classification

For the purposes of this book I will use an alternative system to Singer's classification of ball lightning theories. This is entirely arbitrary but it does reduce the number of categories and classifies a theory based on its main energy source, though some theories overlap different categories. I have chosen five main categories: chemical theories, mechanical theories (including vortex theories), combustion theories, electrical-plasma theories and nuclear theories.

CHEMICAL THEORIES.

Most chemical theories have the advantage that fuel is available to sustain the ball lightning for potentially long time intervals. Another positive feature is that fair weather ball lightning becomes a real possibility because combustion theories are not dependent solely on the existence of lightning from an electrical storm. One apparent disadvantage of these chemical based theories is that the electrical effects in combustion theories are not directly explained. This may not be a problem. I found the number of cases of direct electrical effects associated with the ball lightning object were, in fact, quite small. I will provide a possible explanation of these reported electrical effects later in the book in terms of the vortex fireball theory.

Muschenbroek in 1769 stated that ball lightning is not an electrical phenomenon at all, but an agglomeration of flammable materials falling downwards from the upper atmosphere. In order to get to the upper atmosphere these substances were thought to come from the Earth's interior and vaporize upwards. The flammable substances were then ignited by some unknown mechanism. Muschenbroek's theory appears to be a precursor to many chemical theories of combustion, including my own vortex burner hypothesis. However, the theory was silent on how the fuel assembles itself into a spherical volume and precisely where the fuel comes from.

Barry (1968) theorized that ball lightning originated from the combustion of simple hydrocarbons. In the first step a lightning stroke into a hydrocarbon gas generates more complex and heavier hydrocarbons which are then brought together by an aerosol clumping process. Such an aerosol clumping effect was first observed by Cawood and Patterson (1931). This effect was invoked to explain how the fuel gas was brought into a small volume.

Schonland, a well-known researcher on lightning, hypothesized that ball lightning was the result of the combustion of low concentration gas such as methane. A large lightning strike was thought to release the marsh gas near mountain streams and the gas was then ignited by a subsequent lightning strike. I can see why this theory was not accepted because not only have many cases of ball lightning been seen in non-marsh areas but they have also been observed in fine weather with no association whatsoever with ordinary lightning. The theory could not explain how the methane is assembled into a spherical volume.

Not all ball lightning chemical hypotheses utilize hydrocarbon as a fuel. Smirnov, in a series of papers from 1975 to 1977, thought that ball lightning consisted of a series of reactions between oxygen, ozone and nitrogen dioxide (Charman, 1979). He claimed that with an initial ozone concentration of about 1%, a relatively low temperature of a few hundred degrees Centigrade would be suitable for combustion. The radiation produced by the chemical reactions was estimated last only for a few seconds.

The physical reason why the flammable material should become localized into a small volume in the upper atmosphere is a common problem that plagues many of these flammable material theories. However the theory of Cawood and Patterson is a noticeable exception. In an experiment they produced a non-combusting charged aerosol cloud that clumped together in a quasi-spherical form and could be touched with a probe. No work has been done, as far as I know, on a combusting Cawood and Patterson charged aerosol ball. There are more problems with pure combustion theories, which I will describe later.

Smirnov's fractal theory

Mandelbrot created the mathematical theory of fractals and its associated fractal dimensions. Instead of one, two, or even three dimensions, the theory is able to deal with fractional dimensions eg 2.7 dimensions. The theory has been widely applied from such things as snowflakes to the coastlines on the earth's surface. The theory of fractals has even been applied to ball lightning in the form of a theory of charged aerosol clusters put forward by Smirnov (1987). These aerosols were held together by Van De Waal's forces, which are weak electrostatic forces. The structure so formed, was an "aerogel", or hollow fibrous structure, analogous to balls of eider down feathers from a sleeping bag. This aerogel structure of tiny fibers joined together in a three dimensional array was said to possess electrostatic "tension" in this theory and explained the presumed property of surface tension attributed to ball lightning. The size of the surface tension was calculated to be close to that of water. Chemical oxidation reactions on the aerogel structure were thought to adequately explain light emission from ball lightning.

Smirnov (1990) considered ball lightning to be a fractal phenomenon, but to my knowledge, there is currently no evidence of such an aerogel structure existing in nature in connection with ball lightning. Singer (1997) is of a similar opinion. I have read reports on Russian experiments with exploding wires that claimed evidence of fractal clusters, but it does not follow that fractal structures are the definitive structure of ball lightning.

Smirnov pointed out one advantage of his fractal cluster ball model. His theory was said to have the capability of explaining the conservation of the spherical form of the ball as it passes through small holes. The aerogel structure can compress itself into a smaller volume and squeeze through a small opening.

Smirnov (1994) in his report on two international conferences on ball lightning made the unsupported claim that ball lightning was formed of nanometer-sized particles-from laser experiments which vaporize pieces of metal (usually wire) which then condense into a network of fibers termed an "aerogel". I have found no conclusive evidence in the scientific literature to support the thesis that naturally occurring ball lightning has a fractal nature.

Abrahamson's silicon ball theory

There are many variations on the filament or aerogel theories where the fine filaments can oxidize in a chemical reaction. Perhaps inspired by the shiny silicon deposits seen in fulgerites, John Abrahamson and his former student, James Diness published a silicon fiber theory of ball lightning in Nature (Abrahamson and Diness, 2000). The requirements of the theory are very restrictive. The basic idea is that an ordinary lightning strike hits the soil and reduces any silicon dioxide present to silicon using carbon as the reducing agent. The lightning makes a fulgerite, which is a long glassy hole. During the discharge process, the lightning produces a vortex like a smoke ring produced by a smoker's puff. The silicon oxidizes to silicon dioxide from an oxygen diffusing process into the aerogel producing the light.

The oxide layer was said to slow down the diffusion rate enabling a slow combustion rate to thereby enhance the lifetime of the ball. However, the lifetimes calculated range up to only 30 seconds with the numerical model used. How does such a theory account for long-lived events ranging up to a few hours? Presumably the aerogel would have to be more closely packed with fibers to limit the diffusion of oxygen. Nevertheless there are reports of long-lived balls recorded in the literature that are not associated with such a metallic source. The problem with this model is that all the fuel has to be supplied at the moment of creation. This leads to lower lifetimes, whereas other chemical models, such as an external source of natural gas, are at a clear advantage. Abrahamson (2002) modified his silicon theory to include metal obtained by lightning striking a metallic surface.

Despite the publicity surrounding this theory, when it was published in Nature, the concept does not match many of the reported properties of ball lightning. It is therefore a more incomplete theory. Others have already pointed this out. My thinking suggests the following problems:

1. The fluff ball model requires a lightning strike that does not tie in with fair-weather events not connected with lightning directly.
2. The theory cannot explain ball lightning that can dig trenches or otherwise show large amounts of mechanical energy.
3. The theory does not seem to have the capability of explaining rotation and movement against the wind and other unusual trajectories.
4. Ball lightning has been seen in locations devoid of soil, metal and lightning strikes. For example of the wing tips of planes.
5. Every ball lightning event should have a fulgerite signature. However this is contrary to the evidence since I cannot cite any published event where ball lightning has been directly connected with fulgerites.
6. The theory limits lightning balls to a diameter no greater than 30 centimeters whereas ball lightning can be a few meters in diameter.
7. Under this theory there should be no correlation with fault lines and tectonic activity. Certain lights showing similarities with ball lightning have reported such correlations.
8. Ball lightning exhibits other geometries that are difficult to reconcile with only the smoker's ring vortex.
9. The inability of the theory to account for long-lived events.

These are some of the observational problems facing the Abrahamson-Diness theory. The other difficulty is that despite trying to establish the supposed effect in the laboratory they have been unable to produce anything like a 30 centimeter ball of light. The experiments produced some nanometer structures that were found in the air above the strike. Later experiments produced some promising puffs of smoke, which were conjectured as being precursors. Air was sucked in through a tube, collected by filters, and later analyzed using an electron microscope.

The aerogel concept may have a part to play but it is unlikely to be the main player in understanding ball lightning and associated lights. In the vortex fireball theory fibrous material may carry electrical charge and be contained in the vortex but is only as a secondary effect. Uman, a lightning researcher thinks that the "fluff-ball" model has gaps. He and his associates have been within one hundred meters of lightning strikes to the ground (rocks, soil etc) on many occasions and not one of them has seen a ball lightning event. Nevertheless, one could claim that for ball lightning to be created requires an unusual lightning strike, one that produces a fulgerite. The problem is that plenty of lightning-generated fireballs do not produce fulgerites. Hubler and Uman were also not convinced about the mechanism to bring the charged nanostructure fibers into a ball structure.

ELECTRICAL AND PLASMA THEORIES

Three generic electrical theories of ball lightning

There are three types of electrical theory of ball lightning which have been discussed in the scientific literature. These are the electrostatic arrangement, plasma, and electro-magnetic radiation models. Such theories have been particularly popular with physicists, although the outstanding physicist, Michael Faraday who made fundamental discoveries in the field of electricity and magnetism, thought that ball lightning was not essentially electrical in nature. His view was that reports of ball lightning had a much longer lifetime than the usual electrical spark discharge he was accustomed to.

Electrostatic theories

Many electrostatic theories use the notion of a Leyden jar structure. The Leyden jar was an early and crude form of capacitor, which stored electrical charge as a battery does. It consisted of a glass cylindrical container with an inner silvered surface. The Leyden jar structure, as it applies to ball lightning, would be a quasi-spherical capacitor arrangement. Electrical charge resides on one spherical surface while electric charge of the opposite sign lies on another concentric sphere. The electric field is then radial and intersecting the common center of the two spheres. If there is a sufficient amount of charge existing on these spherical surfaces then the breakdown voltage of air (about 30 kVcm-¹ at 1 atmosphere) will be reached and an electrical spark discharge will then emit light giving ball lightning its luminosity. Without this equilibrium of forces the ball would either explode or implode.

The problem with electrostatic models, such as the Leyden jar structure, is the difficulty that they have in explaining long lifetimes. There must be some internal mechanism to continuously supply charge to the surface of the spheres as the charge recombines.

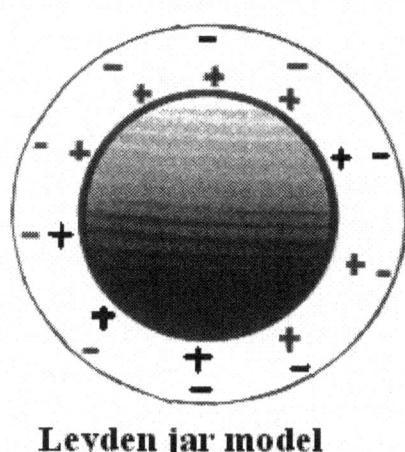

Leyden jar model

Figure 1 : The positive and negative charge separation in the Leyden jar approach to ball lightning.

Singer (1971) cited several dust or droplet ball lightning models. For instance, De Tastes in 1885 described ball lightning as a biochemical process which emitted light when the fermentation of pollen took place. Walker in 1909 proposed a theory of charged water droplets. Puhringer in 1967 described ball lightning as a Leyden jar structure of oppositely charged water droplets.

One recent, yet sophisticated Leyden jar model was advanced by Turner (1998) inspired by the work of Stakhanov. His spherical model has an outer refrigeration zone (to account for some cases of ball lightning lacking heat), a hydration zone and a plasma core. Recombination of electric charge is thought to be slowed by reactions involving the thermodynamics of ions in saturated water vapor. The author commented that he had spent time trying to reproduce the model in the laboratory but there were too many uncertainties to make this a viable proposition. The model would not be capable of explaining the digging of trenches in the ground and would be subject to moderate DC fields from a thundercloud. It therefore could not account for fair-weather fireballs possessing what appears to be an independent motion, even against the wind.

Electrical discharge theories

Electric discharge theories of ball lightning are based on the idea that a strong electric field is needed to produce an electrical discharge. The electric field required to achieve this discharge is termed the "breakdown" field strength. Under this theoretical view the visible effects of the electrical discharge are what the observer sees as "ball lightning". There are severe problems with this approach. The idea may seem right as a candidate for very short-lived events, typically less than one second, but for times greater than this, an alternative mechanism would need to keep operating in order to maintain the electric breakdown field. This requirement for a continuous supply of electrical charge is a fundamental problem confronting all Leyden jar-based theories.

Lightning stroke theories

Singer (1971) briefly described ball lightning theories based on the observation that ball lightning is intimately connected with a lightning stroke. In some theories, a piece of plasma detaches itself from the main stroke and curls up into ball lightning. Mathias, cited by Singer (1971), wrote a number of scientific papers, in which he described ball lightning as being produced after the lightning channel has cooled. The cooling effect somehow served to increase the surface tension of this "fulminating matter". This process is analogous to skin forming on the surface of hot milk as it cools. Apparently, this theory may be able to explain the merging of two ball lightning, just as two water drops are able to coalesce together to form one bigger droplet.

Du Moncel in 1854 hypothesized that ball lightning is a spherical electrical discharge. The discharge was said to spread out because of aerodynamic effects on going through a resistive medium. Phillips in 1888 proposed a theory interpreted in modern lightning research terms, as the luminous front of an advancing leader lightning stroke. Toepler in 1900 published his idea that ball lightning was a segmented part of plasma from a lightning discharge in a lightning channel from a previous stroke.

Theories based on the lightning discharge have several serious shortcomings. The diameter of ball lightning is limited to around the actual lightning channel diameter which contradicts contemporary research demonstrating that ball lightning can have a diameter much greater than this. Motion against the wind is possible in some of these theories, but only through the rather mechanically contrived concept that the breakdown electric field acts like a beam and remains focused and moves against the wind. It is unrealistic to expect the electric field to remain at such extreme breakdown values and to be so well focused, like a beam, to produce a spherical glowing region from an electrical discharge. Then there is the question of how this beam manages to move around in a way that it satisfactorily reproduces the diverse trajectories observed. The trajectories of ball lightning range from a graceful geometrical curve to a regular zigzag motion. One particularly difficult observation for this theory to account for is the motion inside metallic buildings that could shield out any external electric fields.

Plasma theories

Plasma is often called the "fourth state of matter", along with the more familiar states of matter of solid, liquid, and gas. Plasma is a state of matter in which all atoms are totally ionized to produce a swarm of electrically charged ions such as electrons and protons. What plasma theorists have suggested is that the plasma left behind after the passage of a lightning stroke, is the "stuff" of ball lightning. The problem is that very soon after the discharge; the charged ions will recombine in a matter of milliseconds. Positive ions neutralize negative ions and there is a return to neutral matter such as a gas. Hence plasma theories normally predict very short lifetimes for ball lightning, much less than a second. What physicists do is try to imagine a special effect that somehow prevents the "recombination" or "neutralization" of this charge. It is therefore not surprising that theorists have devised imaginative schemes to overcome this problem. The plasma theorists have directed their attention towards constructing a theory that would somehow entrap the plasma for relatively long time spans within a magnetic "bottle".

Plasma physicists hope that by understanding ball lightning they can succeed in their goal of sustained "plasma confinement" over long time spans and at very high temperatures. Plasma confinement is a kind of Holy Grail quest for long-lived nuclear fusion in the laboratory. Prolonged confinement of plasma is necessary at very high temperatures so that hydrogen nuclei can collide together with sufficient kinetic energy to produce helium. Long-lived nuclear reactions like this would mean a cheap and limitless supply of energy.

Silberg in 1965 invoked the electron gas quantum theory of ball lightning. In his theory, confinement of a totally ionized gas in a sphere takes place by the quantum mechanical exchange energy of electrons. The spheres were estimated to have a relatively low temperature of 632 K, but of a high charge density. It is not surprising that such exotic phenomena have not been observed in nature.

Self-contained plasma vortices

Plasma vortex theories were briefly mentioned in Singer (1971). Many of these plasma theories are dependent, in some way, on the existence of either the lightning stroke or the lightning channel moments just after the passage of the stroke. Hence they suffer from the same problems as all lightning-based theories as I mentioned earlier. I will mention a few descriptions to illustrate the diversity of this type of theory. Balyberdin in 1966 theorized that ball lightning was an electron vortex. He was attempting to physical model the formation of ball lightning, on, or near, the nose of an aircraft after a lightning strike.

In 1905 Carlheim-Gylenskold, (quoted by Brand in 1923), proposed that ball lightning was a rotating vortex of ionized air separated from the cylindrical discharge channel of lightning. Wolf in 1915 proposed that ball lightning was an electron vortex ring formed by a pulse from ordinary lightning. In this model electrons were thought to ionize the air by collisions to produce a vacuum inside the ball.

Carpenter in 1963 described ball lightning as originating from the conversion of a plasma rod into a plasmoid with the structure of a "Hill's vortex". A Hill's vortex is easily represented pictorially, and is a theoretical solution to the Navier Stoke's equation for fluid flow. The conceptual model involved the self-confinement of ions and electrons by the trapped magnetic field. This is a magnetohydrodynamic (MHD) model because the ions and electrons of the plasma are held by a magnetic field. The geometric structure was presumed to be a torus, which is a donut shape. Very high currents of 10 million Amperes were assumed in the model, and another extreme condition was that electron velocities were near one third the speed of light. When electrons are going this fast, the rest mass of the electron increases in accordance with Einstein's special relativity equation. This ball lightning theory employed aspects of both quantum mechanics and relativity theory.

Gubichev in 1966 conceived of ball lightning as a positive ion current torus from a shock wave disrupting a linear lightning channel. A shock wave was said to interrupt the current flow and the magnetic field around the channel collapsed at the same time transforming the ionized channel into a vortex.

Plasma vortex theories would appear to face the same sort of problem as non-rotating plasma theories, namely the problem of confining the plasma long enough to prevent recombination. There is also no obvious mechanism that would propel the plasma along the unusual flight paths observed in relation to ball lightning, such as moving against the wind. There is a further problem. If ball lightning was seen outside an airplane window hovering around the wings, such plasma would be blown away and quickly cooled.

One problem confronting ball lightning plasma theorists is to explain the seemingly enormous energy densities often reported. A celebrated example of this is the incident in which a spherical ball lightning, the size of an orange was seen in Dorstone, Hereford on 3rd October 1936. It dropped into a barrel filled with four gallons of water and boiled the water in the barrel for several minutes. Twenty minutes later the water was too hot for the hand. Professor Goodlet (Goodlet, 1937) calculated that the minimum energy of this fireball as 3.8 mega Joules. This report implied a very high energy for something the size of an orange. It is easy then, to see why some plasma theorists have resorted to explaining ball lightning as black holes and antimatter. In my theory there is no need to do this, because the total energy could be easily explained by supposing that chemical energy from the outside is being supplied to the combusting vortex. This intake of fuel gas could sustain an orb over the reported time interval.

Another difficulty for plasma theories is to account for the observation that some ball lightning objects eject fiery debris. The reason is that most plasmoid theories use solely the concept of plasma. In my opinion some extra feature is required of the theory to overcome the difficulty of explaining this throwing out of burning material. On the other hand, such an observation is easily explained using a combusting vortex where material is centrifuged out and then drops to the ground by gravity.

Electromagnetic radiation and microwave theories

One popular idea which has served as a guiding hypothesis for several investigators is the notion that ball lightning is created by electromagnetic radiation from an electrical storm, which is then focused at some spherical region near ground level. An electric field of sufficient strength, and associated with the radiation, is required to ionize the air and form plasma. Such theories are called "natural electromagnetic radiation" theories.

Several such radiation theories, and their variants, exist. Charman believed that credit should be given to the original investigator, Cerillo, who first proposed these electro-magnetic wave theories of ball lightning back in 1943. However, the most well known natural electromagnetic radiation theorist would be the Nobel Prize winner and Russian scientist, Peter Kapitza. His theory suggested that ball lightning was a plasma ball created by an external radio frequency beam with the wavelength of the radiation roughly corresponding to the diameter of the ball (i.e. 10-20 cm). Kapitza (1955) theorized that ball lightning is a plasma created by a standing electromagnetic wave that excites a low concentration of ionized gas from a previous lightning stroke. Absorption of power from the wave is most effective at resonance when maximum power is transferred to the ball. Resonance takes place when the driving frequency of the source equals the natural frequency of the object that is resonating. This storm resonance phenomenon is analogous to the vibrating tuning fork which induces a guitar string tuned to the same frequency, to sympathetically vibrate with strong amplitude. My understanding of Kapitza's theory is that the resonance condition determines both the resonant frequency and wavelength of the "ball", which in turn, also determines the diameter of the ball.

One significant problem, often quoted to refute such radiation theories, as Kapitza's, is that the electric field of the radiation from electrical storms (in the required frequency range) has been measured and found to be several orders of magnitude less than that required to generate an electrical breakdown of the local air.

V.G. Endean's spinning electric dipole theory

Geoffrey Endean proposed a ball lightning theory where microwave plasma confinement is the essential element. I will take a look at his theory in slightly more detail. Dr Endean is a scientist who takes an active interest in the ball lightning problem having turned up to several International Ball Lightning Symposia. In a paper to *Nature* he postulated that ball lightning could be described as spinning electric dipole model, (Endean, 1976) where the spinning electric field vector is in the microwave frequency region.

The Endean analysis is a follow up on a suggestion by Professor Jennison, who actually saw two ball lightning objects in a plane at different times. Professor Jennison suggested that ball lightning is a spherical mode of a standing wave of electromagnetic radiation. The idea was that ball lightning's nature is essentially electromagnetic field energy trapped in a quasi-spherical partial vacuum and an ionized sheath partitioning it from the atmosphere. Endean's theory seems to be related to the work of Dawson and Jones (1969). They conceived ball lightning as a set of microwaves trapped in a bubble of highly ionized, conducting, spherical walls carrying large surface currents. Most of the energy was stored in the microwave radiation, rather than the plasma walls.

Its creator cited two advantages of the spinning dipole theory. One of these is that the theory seems to agree with the variation of energy density variation with a so-called "radius parameter". The graph was reported to be consistent with the energy density range graph of Smirnov (1987). The other evidence was that the theoretical model, in having a zero magnetic field, admits a second order magnetic field around the ball. This second order effect was thought, in practice, to attach a minor role to magnetic field effects. This seems to be true of ball lightning observational data where there is little evidence to support a strong magnetic property of ball lightning.

The spinning dipole theory was not discussed in terms of how it could deal with other physical properties of ball lightning such as its odd motion against the wind, the splitting into two or more balls and then recombining and so on.

Another problem with the spinning dipole theory is the assumption that the ball's creation is directly associated with a lightning strike. Recent research demonstrates that ball lightning has been seen in fair weather. The Japanese, in particular, have collected a number of observations along these lines. As well as this, the spinning dipole theory requires some object in the surroundings, such as a building, or tree to influence the usual path of a lightning stroke. This restrictive asymmetry (lack of symmetry) requirement of the spinning dipole poses a problem. It obviously does not explain the case where ball lightning has been observed far away from any asymmetry in the environment.

In a later paper, Endean (1993) pursued his electric dipole concept and presented an analytical solution, in spherical geometry, of his earlier model published in *Nature*. The rotation rate of the electric field vector was assumed to be in the microwave frequency range. His model also postulated that the total magnetic field was zero. This assumption was said to greatly simplify his mathematical analysis and enabled a general limitation, applicable to plasma theories, called the "virial theorem" to be side stepped. The virial theorem basically restricts the energy density of a self-contained plasma ball. When applied to a sphere, the energy density inside the sphere, where there are only electromagnetic fields and no tensile strength, is equal to three times the average pressure at the surface of the sphere. The virial theorem is well known in the ball lightning field yet several theorists have failed to appreciate the constraint the theorem places on their own ball lightning theories.

Endean concluded that there was evidence for the spinning electric dipole model in nature, but it remains unclear whether this evidence can be shown to be causally linked to his model. The evidence consists of video tape recordings reported in Charman (1979) by Eriksson (1977). I checked out the above two references cited by Endean, and although he says these references support the theory, I could not find anything to satisfactorily substantiate this claim. Eriksson, from the Transvaal Highveld region of Southern Africa, stated that there was some doubt about the authenticity of the images on the tape. The whole event was captured on film but it lasted only about 180 milliseconds and the "ball" did not occur in every frame (nine in total). In fact, I counted four round light forms in frames 2, 4, and 6 and something odd in frame 7.

Mention was made by the author that the blooming of the lenses is a typical feature of this type of camera when lightning strikes close (around 700 meters). Blooming may have taken place and it was said that it could not necessarily be discounted in this sighting. Eriksson, himself, did not regard the sighting as a definitive sighting of ball lightning, but only as an event open to further discussion and interpretation. He suggested that further evidence would be required to settle the case.

Even if the image was positively confirmed as a genuine ball lightning sighting, there is no direct evidence cited by Eriksson that the image was in anyway directly connected with Endean's spinning electric dipole model.

K. Nickel's problem with electrical theories

Nickel (1988) found a significant problem with purely electrical hypotheses, which I believe, is still valid. If ball lightning's nature is essentially electrical, then the ball should short out to earth and lose its charge when it bounces or touches a conductor such as a metal object. A number of reports of ball lightning show that if it were electrical it would have discharged to a conductor. However, no such discharge was observed. Cade and Davis (1969) on page 89 cited a case from Dr Dewar's review of 513 ball lightning reports. A rotating ball of fire like a balloon was seen in about 1939 in Palestine, Illinois. It descended to the floor but it also appeared to bounce lightly. It traveled through a steel wire screen (mesh size unknown) and bounced on a window sill. Surely any purely electric ball would have shorted out when passing through a metal screen and lead to the ball's demise. As far as I know this problem with electrical theories has yet to be adequately addressed.

Lowke et al's metallic vapor theory

One type of theory suggested that ball lightning is a metallic vapor and can reach incandescence and glow like a light bulb. Lowke *et al* (1969) used a mathematical model using a heat equation that assumed a sphere of hot air at initial temperatures of 1000 and 10,000 degrees Kelvin. Cooling rate curves of the hot sphere were calculated. It was concluded that decay times could be in the order of seconds. However there are problems with this model. The visible output intensity decreases rapidly and the hot sphere will always rise because of buoyancy forces. The only way the authors could get radiating spheres to descend vertically was by seeding the hot spherical air region with copper vapor. This would increase the average density of the sphere beyond that of the surrounding air.

The problem with metallic vapor theories is that they appear to have limitations in accounting for the observation of ball lightning moving in a lateral direction against the prevailing wind direction, or the more dramatic changes in motion sometimes observed. The reason is that if ball lightning is a hot buoyant spherical region of gas containing material, such as a glowing metallic vapor, then the motion of such a "ball" is either upwards or downwards depending on the density. In this theory there is no mechanism to produce a horizontal component of motion, in the absence of air, in spite of ball lightning reports which indicate that the movement of ball lightning can sometimes take place in very unusual paths and with diverse sideways motion.

NUCLEAR THEORIES

The publishing of nuclear theories may have been a response to the very high energy densities cited for ball lightning, particularly the classic case of a fireball, the size of an orange, landing in a barrel of water (Goodlet, 1937). Such theories are still being proposed with little thought given to other important properties of ball lightning, as published by ball lightning researchers.

Nuclear theories abound. Bottlinger, as early as 1923, suggested that if very strong electric fields were present in the atmosphere, nuclear reactions could be triggered. The problem is that there is no evidence to suggest such reactions actually take place in nature.

Nuclear theories that invoke nuclear processes would obviously require reactions involving the proton, and the neutron. Nuclear theories also predict an increase in levels of radioactivity and there is no real evidence for this in connection with ball lightning. Dauvillier in 1957 speculated that the energy of ball lightning could be derived from atmospheric nitrogen converting to carbon-14 through a nuclear reaction. Ball lightning would then be composed of carbon-14. Again there is no evidence of such nuclear reactions in nature, and no hard evidence of carbon-14 or increases in radiation levels following the passage of ball lightning.

Arabadji in 1957 proposed that nuclear reactions could take place due to the intense focusing of high energy cosmic rays produced in thunderstorms. However, there is a lack of any evidence for cosmic radiation in association with ball lightning.

Singer (1971) pointed out more difficult problems that nuclear theories of ball lightning would need to overcome. The transfer of energy in an atmospheric process, via, say a lightning stroke, is unlikely to get charged particles reaching sufficiently high voltages that are required for a nuclear reaction. In addition, the maximum temperature observed in lightning strokes is reportedly too low to achieve a thermonuclear reaction. Cosmic ray theories that propose the generation of nuclear reactions at regions near the earth's surface are not realistic. Measured fluxes of cosmic rays in the lower atmosphere region are not sufficiently concentrated to generate nuclear fireballs.

Altschuler *et al's* (1970) contribution to *Nature* attracted plenty of attention at the time. They sought to explain the special cases of high energy ball lighting events and thought that high-energy forms of ball lightning could be radically different to low-energy balls. They presented a self-contained fireball argument that led to a nuclear explanation involving a certain concentration of oxygen-15 and nitrogen-14. But is such an assumption correct? I suggest that it would be more realistic and in line with Occam's razor to find out if the energy could have been supplied to the ball from the surroundings. The vortex burner theory is capable of explaining high-energy events using` this idea. In one spectacular case reported by Botley, a 60 centimeter diameter fireball dug a 100 meter long trench to a depth of 1.2 meters in soft soil by a stream, as well as excavating 25 cubic meters of stream bank. Based on these observations, the energy density was estimated to be more than 10^3 Jcm^{-3}. This trench excavation event demonstrates that ball lightning is a phenomenon with a large amount of mechanical energy. The "energy density" can be easily be explained by a combination of chemical combustion energy and the kinetic energy of the hydrodynamic vortex. The latter is well known to have high mechanical energies and would account for the pattern of damage where a powerful vortex can lift soil and make trenches or holes in the ground.

Nevertheless, Altshuler *et al* dismissed several classes of ball lightning theory with the following critique. They used an energy argument based on energy density and eliminated self-contained chemical theories because the energy density was too low. They assumed a closed system where energy was not supplied from the surroundings but with apparently no justification. They also discounted electrical theories because the recombination time is too short-perhaps a few tenths of a second. They ignored hot plasma theories because these predicted that the ball would rise due to a convective up draught and be disrupted by air flows. They maintained that hot plasma could not be magnetically confined at atmospheric pressure at the high energy density densities required. The authors went on to postulate cooler plasmas which were confined by quantum exchange forces or a superposition of plasma modes which apparently do not produce enough light to form a glowing sphere of ball lightning.

I agree with several ball lightning theorists whose analyses conclude that nuclear theories are unrealistic in being able to explain ball lightning physical properties.

VORTEX THEORIES

The origin of vortex theories quite likely developed from the simple observation that ball lightning has been seen to rotate. Some theorists have proposed that ball lightning is some kind of a rotating mass of air. Faye in

1890 suggested that ball lightning be formed from whirlwinds, cyclones and tornadoes. The basic concept was that ball lightning is a rotating sphere of highly charged water droplets and hail etc and they fireballs detach from the funnel tip. The model was developed after fireballs the size of billiard ball melted 8 centimeter holes in glass windows facing the storm. Such vortices are called "hydrodynamic". The idea that ball lightning is a vortex does have some historical precedence. Hands in 1909 and 1910, and Lagrange in 1910, both thought ball lightning was light from whirlwinds (Singer, 1971).

In nature these free vortices manifest themselves in such forms as dust devils and tornadoes. But there are more complex and interesting forms of vortices, like magnetohydrodynamic vortices that some theorists have used to explain ball lightning.

Problems and advantages of the vortex idea.

Vortex theories have not escaped criticism. I will look at two objections to ball lightning vortex theories. The first is the claim that a vortex would easily dissipate. The other difficulty raised is explaining how ball lightning can go straight through a window like a ghost leaving the window intact. In the latter case such reports of this nature are very rare but it seems to have inspired a whole class of theories involving electromagnetic waves. I believe these problems may be overcome. To say that a vortex can easily dissipate is not really a problem at all. So long as there is both an ample initial supply of vorticity and vertical momentum (the updraft) the vortex will continue. Many ball lightning events might last only a few seconds but this would be ample time for a vortex to exist. The second difficulty cannot be surmounted and a vortex theory does require some opening for a vortex to enter a building. The exception to this would be when a fireball breaks or melts through at a weak point like a glass window, or a vortex could be created by an impact such as a lightning strike on the outside of a wall. This has happened. A fireball manages to enter a room at a meteorological institute in Scotland. The fireball managed to cut a hole with the edges of the glass made smooth by a heating effect. The central area of the glass remained more or less in tact. If there was a vortex fireball it could have punched a hole and seared the edges as it passed through. A slower "ball" might have melted the inner region as well.

The lifetime for a vortex can range from seconds to hours, which is clearly an advantage in explaining both short and long-lived ball lightning. Of course the vortex must be able to generate all the features one normally associates with ball lightning, including its luminosity. Vortex-breakdown is a starting point to explain the spherical form of some ball lightning. Curiously not all ball lightning are spherical. The cylindrical form, for example, could be explained using the tube geometry of a hydrodynamic vortex.

There are many advantages in having a hydrodynamic columnar vortex theory of ball lightning. Such a vortex in the Earth's atmosphere is essentially ubiquitous, and can range from a few centimeters in size to upwards of several meters. Such a size range could easily account for the reported range of ball lightning diameters, which has a similar continuum. There is no need to assume that ball lightning is capable of possessing extremely high energy densities. Such unrealistic theories, like those derived from antimatter, can be avoided. Instead, a fraction of the total energy could be mechanical energy in the form of kinetic energy of the rotating air mass, and chemical energy continuously added to the ball from the surroundings. Further support for the vortex notion is the strong association of vortices with ball lightning reported in the literature, leading more that one scientist to speculate a possible causal connection between ball lightning and hydrodynamic vortices.

Kikuchi (1995), in a contribution to a handbook on atmospheric electrodynamics, completely omitted any reference to ordinary hydrodynamic vortex theories of ball lightning. He stated that ball lightning is one of the biggest mysteries of atmospheric electrical activity. I think the assumption that ball lightning has an intrinsic electrical nature has currently no credible scientific basis. Scientists are still not in agreement as to the true nature of ball lightning and there are many competing hypotheses of ball lightning, not only those that postulate a fundamental electrical nature.

Vortex theories with electrical discharges

Vortex theories that postulate the electrical discharge from small particles inside the vortex are interesting as a model for ball lightning. They suffer the same defects of ordinary electrical theories, like being dependent on lightning discharges under storm conditions. Then there is the possibility that charged aerosol clouds within a vortex could be implicated. Charged aerosols collected by a small atmospheric vortex would not have a sufficient electrical charge density to sustain a continued electrical discharge. I explored this aerosol idea in my thesis and came to conclusion that for a small vortex the size of a typical 15-20 cm diameter "lightning ball", the charge densities would be insufficient to sustain an electrical discharge in a spherical region. Furthermore, recombination of charge would be fast, unless there is a steady stream of charged particles to the vortex. Another problem is that the vortex would need to successfully separate the charge into two airborne "electrodes''.

The electrical vortex hypothesis might seem conceivable if the fieldwork on dust devils is anything to go by. The measurement of high electric field generation near dust devil vortices has been reported. Crozier (1970) reported on a program that measured the electric fields of dust devils in a semi-desert mesa 30 kilometers SE of Socorro, New Mexico. In one dust devil the electrical field was large enough that Idso (1974) actually saw an electrical discharge from the base of a large dust devil towards the ground. This discharge could have come from frictional effects from dust swept off the ground and into the vortex.

I observed this charging effect experimentally in my small-scale vortices. In Coleman (1990), as a working hypothesis, I tentatively proposed a preliminary mechanism where the action of the vortex separates larger, positively charged particles from lighter negatively charged particles. As the vortex lifted these charged particles off the desert floor the vortex took them into the vortex-breakdown recirculation region and formed quasi-spherical particle sheaths. I imagined that these sheaths could act as electrodes. Discharges from one sheath to another may then produce the luminosity of ball lightning. There was support for the existence of these sheaths in nature. Dust sheaths have been observed in dust devils (Snow 1984). However, I moved away from the idea of an electrical vortex theory for ball lightning because of the above problems, and because I found the vortex-breakdown combustion theory better placed to overcome the problems that have plagued other theories.

Vortex theories that require storm conditions

Singer's review contains several vortex theories that use lightning as the initiator of the vortex. Moigno in 1959 hypothesized that ball lightning resulted from the collision of oppositely-directed lightning strokes producing a glowing sphere. Meister, in 1930, described how ball lightning might occur from a sharp change in the direction of the lightning discharge channel, causing a parallel, but opposing air flow which generated a rotating vortex. The result was a rotating layer of air surrounding with an evacuated inside. The centrifugal force arising from the vortex balanced the force from atmospheric pressure. Flint in 1939, followed by Gold in 1952, both described ball lightning as resulting from a descending lightning leader carrying a negative charge, which then encounters a positive "streamer" discharge rising up from the ground to produce a vortex.

Some vortex theories directly incorporate electrically-charged dust or droplets in the vortex. For instance, Faye in 1890 viewed ball lightning as a small vortex where electric charge was generated from high velocity frictional contact between water droplets, hail and other solids. Frenkel, in 1940, proposed a spherical vortex containing a mixture of small cloud droplets or dust. The type of vortex proposed, was the so-called "Hill's hydrodynamic vortex", which was suggested as being produced by a lightning strike. In this conception there exist alternate layers of positive and negative charge with air between and charges compressed by the magnetic fields associated with rapid movement of the charge in the layers.

Magnetohydrodynamic vortices

Magnetohydrodynamic theories of ball lightning describe a vortex having an additional magnetic force added to the concept of the hydrodynamic vortex. What we have is not only the mechanical vortex resulting from the gradient of mechanical pressure, but a flow of plasma from a gradient of magnetic pressure. It is easy to understand the gradient idea by using the picture of wind flowing from high to low pressure. However when both the electric and magnetic forces are included the result is the so-called "electro-magneto-hydrodynamic" vortices or "EMHD" vortices. Kikuchi (1991) applied his theory of EMHD vortices to the controversial corn circle phenomenon in England.

Vortex combustion theories

After the vortex-breakdown burner theory was published in 1993 I discovered two other theories that combined both ideas of the vortex and the combustion of a fuel gas. But these theories did not incorporate vortex-breakdown and did not attempt to explain a much wider set of observations. Nickel (1988) wrote a paper for the First International Symposium on Ball lightning. He mathematically described ball lightning using a Hill's vortex where the vortex was a toroidal structure, as in a smoke ring puffed into the air. As I said earlier, Hill's vortex does not exist in nature but represents a theoretical solution to the Navier Stoke's equation, which is a well-known equation in fluid dynamics.

The velocity flow field of the smoke ring was thought to be similar to the expected air flow pattern in ball lightning. The vortex was thought to be generated from a jet of air (say 10 ms-^{1} to 100 ms-^{1}) spawned by the return stroke of lightning. This downward jet moves towards the ground and then curls up to form a Hill's vortex. The situation is analogous to a smoker pursing his or her lips and puffing out a jet of smoke and the smoke ring floats upwards. This theory is limited to lightning strokes that can create the smoke ring structure. For the theory to be viable there must be at least one mechanism to produce big smoke rings to account for the larger diameter ball lightning events. It is much more likely that columnar vortices, like the dust devil, whirlwind or tornado are implicated.

A brief mention of a vortex combustion theory was in the popular account of Cade and Davis (1967), who reported on the work by F.Werner. The authors gave a brief reference to the theory. I have never read any detailed account of the theory and unfortunately Cade and Davis did not give the name or the year of the publication of Werner's theory. Werner was said to have conceived of ball lightning as a vortex of burning gas. Using a smoke ring generator he generated vortices of burning gas and the luminous vortices were reported to have moved over a distance of about fifteen feet. A decomposing log was considered a typical source of the combustible gas.

The above two theories of Nickel and Werner both used a smoke ring vortex as their structure. Such theories are limited in what they can explain in regard to published ball lightning properties. They could not be applied as a unified theory to account for the anomalous fireball events described in this book. But theories based on the more common atmospheric vortices like small whirlwinds, dust devils and the tornadoes would be more realistic and pervasive than specifically the smoke ring type structure.

MORE THEORIES OF BALL LIGHTNING

In my opinion an "exotic" ball lightning theory is a theory that postulates the existence of a rare phenomenon within the field of Physics for which there is little or no evidence and for which a much simpler explanation would suffice (again Occam's razor). In many of these exotic theories there is no evidence from nature to back such a theory. Into this category I would put black holes and anti-matter theories. These theories appear to have two basic premises. The first premise is that the ball lightning phenomenon is an extremely rare event. The second premise is that such events should possess high energy densities. Several types of exotic theories of ball lightning have been proposed. Ashby and Whitehead (1971) postulated that antimatter meteorites originating from outer space could cause ball lightning. Altshuler, House and Hildner (1970) proposed that ball lightning is a nuclear phenomenon.

The production of exotic theories continues unabated. I quote from a report on the 4th International Symposium on Ball Lighting, held at the University of Kent, England in 1995. Dijkhuis (1995) in the 8th of August, International Committee on Ball Lightning (ICBL) Newsletter wrote:

"The morning session of the last symposium day extended ball lightning theory into the high energy realm, with A. Bergstrom as chairman. A microwave solution including non-linear hydrodynamics arrives at a hollow sphere with nuclear fusion potential (A.P. Veduta, rs). A solitary wave equation forms ball lightning from atmospheric ions (B.K.Xu). Cyclic neutron detachment from atmospheric nuclei by high energy electrons in linear lightning with solar activity (Y.Z. Volotkin, rs). A luminous halo marks ball lightning as a natural nuclear reactor with deuterium fusion reactions catalyzed by a toroidal layer of relativistic electrons circulating around a magnetic field ring (A.N. Vlasov). Ball lightning may indicate primordial magnetic monopoles consuming paramagnetic dust particles in the atmosphere (B.U. Rodionov, rd). Elementary particle architecture links

ball lightning with microscopic black holes (L. Vuyk)."

The Fourth International Symposium on Ball lightning at Kent did not publish any proceedings so I am unable to describe in detail any of these theories. However the Fifth International Symposium did produce copies of the proceedings (I donated a copy to the Physical Sciences Library at the Canterbury University as a reference). There were 65 contributions published, including ball lightning sightings, ball lightning experiments and several theories. Over half of the contributions were Russian. Their inspiration for the ball lightning problem could have come from scientists, like Peter Kapitza, who was well known in ball lightning research and who received a Nobel Prize for work in superconductivity. Perhaps the plasma chemist, Smirnov, with his fractal cluster idea might have catalyzed this prolific output of Russian theories.

Handel's maser caviton theory

At the Fifth International Symposium a few new ball lightning theories were proposed and some established theories were further promoted with modifications. I found most of these proposals quite limited in their ability to explain the published physical and behavioral characteristics of ball lightning. Ball lightning theorists as a rule tend to treat ball lightning problem as a special category of their own specific field of research, with little regard to the observational data on ball lightning. They have only to study reliable sources, such as Singer (1971). Take the maser-caviton theory of Handel (1989). His theory relies on a lightning strike creating a population inversion of the rotational energy level of water molecules present in air in concentrations of hundred to a billion molecules per cubic meter. Because the lightning generates a short-lived electric field pulse, this excites a population of water molecules into "maser" (microwave amplification by the stimulated emission of radiation) activity. The maser then creates plasma in a region near the ground at the antinode of the standing wave generated by the maser. When the local plasma frequency matches the maser frequency with a corresponding enhancement of the electric field, a so-called "pondermotive" force pumps out ions from the high field region which then forms a resonant cavity in the plasma. This in turn creates a "plasma soliton" or a "high pressure caviton".

The caviton is represented mathematically as an eigenfunction of the non-linear Schrödinger equation. The Schrödinger equation is normally associated with the bound states of atoms in quantum mechanics. Physically the caviton is thought of as an almost totally empty sphere containing a resonant electric field where ions are trapped by an electromagnetic (EM) wave surrounded by plasma. Thus the maser is not necessary once the ions have been held by the EM wave.

Of course this caviton entity has to have some kind of stability. The stability of such an object has been proven mathematically by Liedtke and Spatschek (1984). However the theory predicts that the stability of these cavitons only takes place if the caviton is above a certain size. Ball lightning with a diameter less than this minimum is not possible under this theory while a wide range of sizes are permissible under the vortex burner theory.

Handel made a few predictions that conflict with the observed behavior of ball lightning. His theory predicts no explosions of ball lightning in confined spaces. Another prediction is that ball lightning should not occur in the mountains because of unfavorable geometry. Singer (1971) cited both cases of exploding ball lightning and its presence in the mountains.

The plasmon theory

Ohtsuki and his colleagues have put forward a theory involving the idea of a "plasmon" generated by earth tremors. The plasmon is a type of wave that propagates through rocks to the earth's surface in the vicinity of tectonic activity. At the Earth's surface this plasmon somehow generates an electromagnetic wave (Ohtsuki and Kamagawa, 1997). What they have done is to connect the luminous phenomena, seen accompanying earthquakes, (known as "earthquake lights") with electromagnetic waves produced by the earthquake. There seems little doubt that electromagnetic waves are being produced, since they have been recorded. However, there is no direct hard evidence that luminosity can be caused by these natural electromagnetic waves. The light intensity, on the other hand, could be the result of light emission from the combustion of natural gas, such as methane released though cracks both before and after an earthquake.

Singer's selection of ball lightning theories

Singer (1997) described four types of theories as exemplar theories being currently researched. He did not endorse any particular theory in his paper, though he did point out what he saw as problems facing each theory. The vortex burner theory was cited as an example of a class of theory involving a vortex either incorporating electric charge or a combustible gas (Nickel, 1989; Coleman, 1993). The second class of theory was an electrical theory with electric charge inside a sphere or torus plasma structure. The third proposal was the aerogel structure (Aleksandrov *et al* (1982) and Smirnov (1993). The fourth theory was the microwave maser theory proposed by theorists like Handel (1989).

Overcoming perceived problems with the vortex burner theory

Dr Singer is well versed in ball lightning literature and able to critique the diverse theories available. He wrote the monograph *"The Nature of Ball Lightning"* which is recognized as a classic in ball lightning research. I will address two of his concerns as they relate specifically to my vortex burner theory. Despite these objections I felt he approved the basic idea behind the theory and referred to it in his paper to the Fifth International Symposium on Ball Lightning. His first concern was that a gas burning vortex would require concentrated methane from places like marshes and swamps. Apparently he thought that the atmospheric concentrations in other locations would be insufficient for combustion. Contrary to this view there do exist many sites around the world where there is natural gas outlets as I will demonstrate later in the book.

His other objection is the idea that a burning gas vortex is unable to penetrate glass windows. Admittedly it would appear to be a difficult proposition for a pure hydrodynamic vortex to achieve but the penetration of a window could take place melting a glass window pane by a strong vortex burning natural gas. It is on record that a natural gas and air flame can have maximum flame temperatures up to around 2000 degrees Celsius. In fact, the temperature of such a combustion process would only have to exceed around 1000 degrees Celsius, which is the approximate melting point of ordinary glass. It is conceivable that a burning vortex, possessing a sufficiently large angular momentum could melt of knock its way through a window, into a room, avoiding a total dissipation of the rotating air mass. This is all the more likely, especially if the swirling flame could quickly burn through the glass and allows the vortex its passage through the hole so produced.

Such a mechanism may explain the large number of luminous globes from a tornado resembling ball lightning. The French Academy of Sciences discussed this sighting in 1890. These ball lighting objects were seen in connection with a tornado. They bored circular holes in windows while others managed to get into chimneys (Singer, 1971). The ball lightning connection with the tornado is not merely coincidental since these ball lightning objects are, in reality, small burning vortices spawned from the main tornado.

3
Solution to the ball lightning puzzle

"It is well worth remarking, however, that an understanding of ball lightning may well be necessary if the tornado puzzle is to be solved." Vonnegut (1960).

I will briefly retrace the steps leading to my solution to the ball lightning enigma. My first encounter with peculiar fireballs began with the study of ball lightning through student research at Canterbury University. The lecturers invited degree candidates to select a two month research topic. Project descriptions from the lecturers were posted on a wall in the Physics Honors room. I was attracted to the following description, written by Dr Von Biel, a specialist on the ionosphere:

Atmospheric electricity.
  "Anybody who has been exposed to the most fundamental study of electrostatics will know that lightning is an electrical phenomenon caused by high electric fields which ionize the air and cause it to be a good conductor. Not so well known is that the earth's surface carries a (negative) charge density and therefore electric lines of force terminate on these charges. Add to this the fact that rain drops and dust particles generally carry electric charges and we have the interaction of moving charged bodies and an electric field. If nothing else we would expect the atmospheric electric field during a rainstorm or heavy winds to differ substantially from the fair weather field.
A Thumbnail Sketch of the Project.
1. Conduct a library search of articles and books on atmospheric electricity. A good place to start is with J.A. Chalmer's book on the subject (QC961.C438).
2. Devise an experiment to measure the earth's electric field at, say, 2m above earth.
3. Construct a simple apparatus as required and make measurements.
4. Report your findings."

  Dr Von Biel's original electric field project of building a field mill and monitoring the local electric field was too straightforward. I discussed possible variations of the fair-weather electric field theme-more extreme cases of lightning. I spouted forth the words ball lightning but I knew next to nothing. I was fascinated with the idea that no one had cracked the problem and I had as good a chance as anyone else. The only knowledge I had of ball lightning was the rumor of its creation by the eccentric Croatian engineer-scientist, Nikolai Tesla at his Colorado Springs facility. The connection of ball lightning with Tesla held my interest since Tesla was an enigmatic and elusive character, yet he made major contributions to science and engineering. He claimed to have produced electric fireballs;

"I have succeeded in determining the mode of their formation and producing them artificially."

[Nikola Tesla from Electrical World and Engineer, March 5, 1904.]

There are scientists even today who are busy trying to build high voltage coils to reproduce ball lightning using Tesla's notes.

Once Dr Von Biel agreed to my proposal I could pursue investigating ball lightning. The impending sense of fun with such an open problem kept me highly motivated. I went to find the research co-coordinator, Dr Valda McCann. She was at the point of printing off the final list of students and their projects. I was politely informed that I had not received my first choice. I was initially disappointed but I did not give up. After a brief discussion, Dr McCann allowed me to proceed. I then spent the next two months fully absorbed in my *"Kugelblitz"* investigation. Through this research I became acquainted with a wide range of ball lightning theories. What really excited me were the physical descriptions of actual fireball events that appeared to provide vague clues but just within reach.

Contrary to what I first expected I found serious scientific books and papers on the ball lightning problem. I starting reading the work of other investigators describing their theories and experiments. This literature provided a much-needed platform to search for a new theory. However it was the vivid descriptions from actual observers of ball lightning that really helped out. This literature supplied the evidence that ball lightning had a distinct "personality". Ball lightning was definitely far more robust than a passive glowing sphere. It was more like some kind of energetic rotating fireball. I decided to base my search for a solution to ball lightning on eyewitness accounts rather that what other scientists imagined ball lightning to be.

Stanley Singer and James Barry wrote influential academic monologues on ball lightning. Singer's book, *"The Nature of Ball lightning"* (Singer, 1971) was an essential tool in the early research phase to critically review the general classes of ball lightning theory. Singer's book provided a convincing critique of the various theories available and helped me to eliminate unpromising theories.

Barry's book, *"Ball Lightning and Bead Lightning"* contains a selection of experimental attempts to recreate ball lightning in the laboratory. He described his own experiment that consisted of a high voltage spark which discharged across two copper terminals inside a plastic container filled with a propane-air mix. Somehow he managed to produce an erratic yellow-green fireball that was a mystery as to what this entity was in his chamber. I thought that if I followed Barry's instructions in his paper I might be able to replicate his experiment Reproducibility is a fundamental standard in science. If something cannot be replicated there will definitely be doubt surrounding the method of creation. On the other hand if ball lightning could be produced then it could be studied more closely.

I decided to repeat Barry's experiment in the high voltage laboratory within the Electrical Engineering Department of the University of Canterbury. I used a series of high voltage capacitors discharging from two copper electrodes into a Perspex box containing 1.4% to 1.8% propane (Coleman, 1988b). I aimed to create the phenomenon but after a few discharges to the Perspex box I was unable to replicate Barry's fireball. Was there some "tweaking" factor or factors that brought the fireball into existence? Barry's experimental fireball was an enigma. I suspected that his fireball might have used the chamber's fuel gas but I was completely uncertain about the role of the high voltage discharge in producing the fireball. Since I was unable to generate any fireball, I concluded that the experiment of Barry may not be reproducible. Although these high voltage experiments were promising, I was skeptical of chemical combustion theories of ball lightning.

A third publication now exists that reviews the whole ball lightning phenomenon. Mark Stenhoff wrote *"Ball Lightning-an unsolved problem"*. The book was promoted by the publishers as superceding previous reviews and surveyed a range of ball lightning theories (including my theory). On the website of the Dutch publisher, Kluwer quoted Stenhoff as concluding that ball lightning is a low energy phenomenon:

"Ball lightning is probably not hazardous, but may be a precursor of ordinary lightning. The conclusion is that ball lightning has lower energy than generally assumed and that some theories are thus redundant. After critically reviewing the current theory, recommendations are provided for future research. This specialist book draws new conclusions about the characteristics of ball lightning and relates these to current theories

This conclusion does not accord with the evidence from published observations. Ball lightning does possess high energy and is very hazardous. It has been known to dig trenches, remove roofs from houses, and even boil water in a barrel for several minutes. Ball lightning theories that postulate higher energies are not necessarily redundant at all, so this low energy approach is a puzzling stance. Stenhoff also suggested that ball lightning is an electrical precursor to lightning. If we do not know what ball lightning is how can we be so sure it is electrical in origin? Nonetheless his review is a valuable contribution to the field and brings together several publications for the ball lightning researcher.

There are other textbooks on ball lightning like the monograph by the Russian scientist, I.P. Stakhanov, entitled *"On the Physical Nature of Ball Lightning"*. This second edition was published in 1985.

The study of ball lightning, in contrast to UFOs, is more and more regarded as an authentic field of scientific inquiry but fortunately ball lightning is widely conceived of as a meteorological phenomenon wholly connected with lightning. Early ideas of ball lightning encouraged this view which is still widely held today. The typical conception is a small fireball (around 30 centimeters in diameter), seen in a lightning storm. It is natural for some people to suggest that ball lightning has nothing whatsoever to do with UFOs. However the contemporary ball lightning research suggests otherwise. Modern ball lightning research clearly shows the ball lightning diameter can range up to several meters and there are reports of it being seen in fine weather. Indeed, Ofuruton and Ohtsuki (1988) reported that of the ball lightning events they reported in Japan, 89.7% were actually seen in fine or cloudy weather. A handful of UFO authors have distinguished between the smaller ball lightning, seen in conjunction with lightning, and a much larger ball of light (BOL) with a diameter of several meters. The BOLs were considered distinct from ball lightning

Although past research had failed to uncover the true nature of ball lightning a more comprehensive range of properties and behavior has been revealed. There is no question that the current scientific status of ball lightning is more diverse than commonly understood. Ball lightning is as unexplained luminous entity, sometimes of a spherical form (but not always), observed in the earth's lower atmosphere, yet possessing a surprising range of movement, size, energy as well as other physical properties.

The ball lightning literature of the scientific tradition contained many eyewitness reports. Fireballs have been seen within buildings and even inside aircraft. Ball lightning can very energetic and reports demonstrate ball lightning's remarkable ability to dig trenches and lift roofs. Various shapes have been reported including: spherical, oval, pear-shaped, torus or donut, disc-shaped, cylindrical, spherical with a spiral tail, and a ball shape with two protuberances on top-like horns. The range of motion is also diverse ranging from: standing still, zigzag, curved motion, unpredictable changes in motion, including against the prevailing winds. As for the color of ball lightning, every hue of the visible spectrum has been reported. Even so-called "black" ball lightning has been seen.

One typical case of "classical" ball lightning sighting was presented in the book by Cade and Davis (1981). They reported on an industrial physicist who saw a fireball move down Fullham Palace Road in London. The ball lightning sighting took place during the summer of 1924 or 1925. The witness was fourteen or fifteen years old when he and his father saw a ball lightning the size of a basketball ball, with a bluish-green color, fall from a cloud. It then moved down the middle of the road for a hundred meters or so. It shifted quickly to one side of the road and suddenly exploded violently with a white flash. The ball exploded and left a gaping hole in the end wall of a row of terraced houses. Clearly ball lightning is not a wholly low energy phenomenon.

During the *Kugelblitz* physics project investigation I came across a tantalizing association of ball lightning with the observations of unusual globular lights in tornadoes by the atmospheric scientist, Dr Vonnegut (Vonnegut and Weyer, 1966). These tornado lights seemed to hold the key to explaining ball lightning. But how was a tornado in nature able to generate the required spherical geometry required of ball lightning? While I was in the Physical Sciences Library at Canterbury University I came across a black and white photograph in a book by J. Eagleman, called *"Thunderstorms, Tornadoes and Building Damage"*, published by Heath of Lexington in the US. The photograph showed a type of spherical compressible vortex in the vicinity of the wing of an aircraft made visible using a smoke tracer. The tracer indicated the air motion to be a helical path on a spherical surface and it seemed to me to be the type of semi-spherical vortex I was looking for.

At that stage I could not exactly see how this aircraft vortex represented the solution. Nevertheless I was inspired to write a rough draft paper entitled *"A hydrodynamic capacitor description of ball lightning and other glowing objects"* to consolidate my ideas. I asked Professor Stedman from the Physics and Astronomy Department of the University of Canterbury to critique this draft. A faint ray of hope- he said it contained a "germ" of an idea. So my first working hypothesis became that of a spherical vortex, made luminous, perhaps by some electrical mechanism.

In 1988 I discovered a book by Jenny Randles called, *"The UFO Conspiracy-The First Forty Years"*. I was astonished. I knew people had talked about UFOs but I had always imagined that UFO sightings were some kind of alien flying saucer-driven by the proverbial little green men. On the contrary, there was no such description but the UFO descriptions bore a striking similarity to the ball lightning reports I had studied for my Physics research report. The spherical fireball shape was a common theme, and the dynamic movement was similar to the behavior of ball lightning. This connection appeared to be an obvious and exciting lead. I became convinced that if an adequate theory of ball lightning could be established then many of the unexplained lights connected with UFOs might also be accounted for. The theory would then be a unifying hypothesis valid for both observations of ball lightning and a certain kind of UFO.

By the end of the physics research project I had a single idea that ball lightning was a type of air vortex and through some mechanism the vortex was able to generate its own light. The first mechanism that I thought responsible was a vortex charged dust model inspired, not only by Vonnegut's tornado descriptions, but also by an observation recorded in an old Russian publication of dust storms in which fireballs appeared on cow horns. Once the three-month physics investigation ended there appeared no further opportunity to explore the luminous vortex idea. But there was still the final thesis year to be done in the Physics Department. I thought I could use this time to further my quest. There was one problem. I knew no Physics lecturer would be available to take on ball lightning as an extended research project.

My final hope was the Engineering Faculty within the University. I had a simple plan of an experiment to produce vortices in a wind tunnel and introduce tracers into the vortex. I went over to the Mechanical Engineering Department to investigate the possibility of conducting the experiment. They referred me to a Professor Williamson. While I walked along the corridor of the Chemical and Process Engineering Department seeking this person, I saw a lecturer sitting at his desk.-Dr John Abrahamson. He was the one person whose expertise matched what I was trying to achieve. I spent the afternoon talking to him about my ideas of a luminous vortex. Of course he assumed that the spherical vortex I was talking about was "vortex-breakdown" because he was so familiar with the phenomenon in his work with cyclones. This was an important step in the research process. It became clear that vortex-breakdown was the holy grail of ball lightning-the spherical vortex I was looking for.

In John's field vortex-breakdown is exploited to increase the efficiency of cyclones that are used to separate particulate material of different sizes. I now had a name to put to my spherical vortex. However the most important practical aspect was that I could now study the vortex-breakdown phenomenon. It was well documented in the scientific literature. But a question remained: why was vortex-breakdown so important?

I now had the opportunity to write a proposal to the Physics Department and conduct an extended experimental investigation into the ball lightning phenomenon. This work was reported in 1999 as a physics thesis, entitled "*Ball Lightning*". The thesis work explored the tentative hypothesis that electrical discharges from dust or ions within vortex-breakdown are the stuff of ball lightning. After the thesis I became convinced that the electrical spark discharge was not the primary mechanism to generate luminosity in ball lightning. On the other hand I was sure the hydrodynamic properties of vortex-breakdown could explain the mechanical effects and motion of both ball lightning and UFOs. This was the very first time that the vortex-breakdown hypothesis had been proposed for this type of fireball in a piece of independent university research.

After finishing my Physics thesis I put the problem of ball lightning on the back burner and turned my thoughts turned towards a completely different research topic. I investigated debris flow surges from 1991 to 1995 within the Natural Resources Engineering Department at Lincoln University, under Dr Tim Davies.

It was at Lincoln University while I worked on the debris flow problem that I found a plausible mechanism for the luminosity in the vortex-breakdown zone. I had become dissatisfied with the earlier null results from the ball lightning thesis experiments so I began a thorough search for a viable solution. Surely there must be a better way to generate emitted light than by charged dust or ionic discharges. The crucial insight took place while I was browsing through a book on swirl burners, under the "TD" classified books on the third floor of the Canterbury University Engineering library. The thought suddenly seized me. Why not combustion? If industrial burner vortices can burn a fuel gas why can't an atmospheric vortex like a whirlwind or tornado do the same thing? I imagined a large whirlwind consuming a combustible gas inside the globular furnace of vortex-breakdown where natural gas is sucked into the fierce heat within vortex-breakdown. Immediately I could see a solution to the whole UFO riddle as well as the ball lightning mystery. The classic 30-centimeter diameter ball lightning could be explained simply as a scaled-down version of the much larger whirlwind. I pondered on why no one had ever thought of the possibility of these atmospheric burners.

I wrote up my ball lightning hypothesis in a letter which was first published in *Weather,* a Royal Meteorological Society of the UK publication (Coleman, 1993a). I called the hypothesis the *"Vortex Burner Hypothesis"*. This hypothesis is the central thesis of this book-the idea that vortex-breakdown acts as the combustion chamber to burn a fuel gas. The picture is then of an aerial "crucible" –a veritable "chariot of fire". When an observer sees this type of UFO they are in effect seeing the luminous glow from the vortex burning a fuel gas. This conception allows for the possibility of secondary features, such as the containment of charged fibers within the vortex. This theory could easily explain secondary things such as the ejection of residue from the ball and the electrostatic properties of ball lightning sometimes report.

I return now to the main clue to solving the ball lightning enigma.

This is the unusual luminous effect in tornadoes reported by the atmospheric scientist, Bernard Vonnegut. There appeared to be some unknown casual connection between tornadoes and some forms of ball lightning. Indeed Vonnegut (1960), in a paper entitled, *"Electrical Theory of Tornadoes"* wrote:

"In addition to the fairly well understood primary and secondary electrical effects discussed above, accounts of tornadoes rather frequently include mention of 'beaded lightning' and glowing or exploding fireballs [Flammarion, 1873]. These phenomena are apparently the same as or closely related to the controversial 'ball lightning' [Goodlet, 1937; Kapitza, 1955] whose existence and nature are still debated. In view of our present almost complete ignorance, we shall make no attempt to discuss this class of observations. It is well worth remarking, however, that an understanding of ball lightning may well be necessary if the tornado puzzle is to be solved."

During the earlier stages I studied the work of Vonnegut, especially his 1960 paper, where he set about explaining the unusual lights in tornadoes in terms of an electrical theory. I read that paper several times over trying to find a lead but I needed another piece of the jigsaw puzzle. My aim was to identify what was causing the luminosity in tornadoes. Vonnegut believed that the puzzling lights in tornadoes might be explained by a better understanding of ball lightning. If I found a mechanism to make a tornado luminous then this mechanism could also account for smaller luminous vortices such as ball lightning. The properties of a small atmospheric vortex could explain the reported mechanical properties and behavior of ball lightning.

FIREBALLS IN TORNADOES.

In my preliminary exploration I came across startling references to balls of fire embedded in the main funnel of a tornado and also references to fireballs being excreted out the bottom of a tornado spout (Singer 1971). Botley (1966) and Bonacina (1946), described how a Puritan Church on the 27 October 1638, was destroyed by an intense thunderstorm in which a tornado descended from the sky, ripped off the roof and lifted blocks of stone from the tower. A fireball was observed and depicted in a drawing, at about that time, as a demon's head coming from out of the base of the tornado. In both Botley and Bonacina's accounts there is no explicit interpretation of this "demon head" at the bottom of the tornado as being ball lightning, in the modern sense, but the description seemed to indicate that indeed it was. In a related event Botley cited M. Jean Dessens from the Central Atmospheric Research Institute. He reported on tornadoes that were able to "vomit" balls of fire from the base of the tornado. Botley stated ball lightning might eventually be understood in terms of a vortex such as the tornado. One witness Mrs L.B. Highlet saw orange balls of fire emitted from a tornado that then rolled down a park.

Meaden's (1989) book, page 49, reported interesting sightings involving balls of fire located on the tornado column. John Braithwaite saw a tornado move towards his house manifesting itself as a large fireball. The Newbottle tornado of 30 September 1872 had an enormous rotating fiery orb that moved six to eight feet above the ground. In Pendeen, Cornwall, on the 25th November 1938, a waterspout was sighted and had a fire actually burning within the waterspout which ejected red sparks. I will discuss waterspouts in a later chapter. Electrical or plasma theories of waterspouts have problems with an aqueous medium because of electrical shorting and rapid cool down.

It would appear that references to "balls of fire" could be a reasonably frequent occurrence in relation to luminous effects of tornadoes. The "ball of fire" description strongly suggested not a phenomenon with an electrical origin, but a quasi-spherical combustion flame that quite literally was a fireball.

THE UFO LINK

While I was absorbed with the ball lightning problem, I discovered another puzzle far better known than the ball lightning conundrum, and even more controversial- the subject of unidentified flying objects or "UFOs".

The very first time I had even read anything on UFOs was while I was browsing the Canterbury University Bookshop in mid-1988. This was during my *Kugelblitz* investigation. I happen to stumble on a book called *"The UFO Conspiracy -The First Forty Years"*, written by Jenny Randles. I was surprised to find that as a general rule there was an absence of descriptions of UFOs as metal alien space ships. UFO descriptions consisted mostly of luminous geometries such as the sphere, torus, or the glowing cylinder. These shapes lacked the detail one might expect if such lights were fundamentally alien spacecraft.

What I found was that these UFO cases were uncannily similar to the ball lightning observations I was studying for my *Kugelblitz* report. Could these sorts of UFO sightings, in reality, be a natural phenomenon and explained using a luminous vortex explanation? Is there an explanation of these UFOs in terms of an unknown meteorological atmospheric phenomenon?

Ball lightning-UFO link in the Christchurch Star

I walked up to the *Christchurch Star* newspaper building and had an interview with a journalist, named Greg Ansley. This was 1988, a key year for UFO sightings. I was prompted to act following Greg's latest column about the 1909 wave of mystery airships in New Zealand that was entitled *"Another flashpoint in Kiwi history"*. These so-called cigar-shaped "airships" could have been identified as luminous vortices. I had no understanding, then, of what caused the light emission from the vortex although the vortex behavior I could interpret a little. I was of the opinion that a columnar vortex was implicated and perhaps some electrical mechanism might be causing the light emission in a vortex. I discussed with him my idea, and he wrote up an article using material from my physics report, entitled *"Kugelblitz"*.

UFOs was a hot topic in 1988 since this was the year that the Knowle's family saw a UFO chase their car while they were traveling through the Nullabor Plains in Australia. The event had a significant media impact here in New Zealand. I suggested to Greg that the theory might apply to this event and the 1978 Kaikoura sighting which sent a media frenzy throughout the country. The published article, called *"Mysteries seen in a different light"* was published in the *Christchurch Star* (Ansley, 1988).

From 1988 onwards I assumed that several UFO accounts in books like Jenny Randle's could be explained as naturally-occurring luminous vortices. However I was still left with trying to understand how an atmospheric vortex in nature was capable of both producing a spherical shape and generating luminosity. These two important questions were essential in establishing a solid foundation for the luminous vortex idea.

DUST DEVIL EXPERIMENTATION

I investigated another phenomenon to see if it had any bearing on the problem of luminous vortices. It has been established that a dust devil can generate strong electrical fields. Is an electrical discharge in a vortex the primary cause of luminosity? I needed to investigate the electrical-charging mechanism in a dust devil-like vortex. Although there had been some reports of strong electric fields connected with dust devils, I found very little direct evidence from eyewitness accounts that unequivocally pointed to an electrical mechanism as generating the luminous glow of ball lightning.

As I said one of my first ideas was of an electrical charging mechanism operating in a dust-devil vortex to assemble oppositely charged particles in the form of an "electrode" within the vortex. The basic idea is the centrifuging behavior of a vortex in which large particles, say positively charged, are shifted more or less in a radial direction out of the vortex more than smaller negatively charged particles. The lighter particles keep closer to the axis of rotation of the vortex core. The drag force of the air balancing the centrifugal force directed radially outwards. Particles are entrained in the vortex in the form of oppositely charged inner and outer sheaths. This required a bimodal (double peaked) grain size distribution for the particulate material on the ground, which seems unnecessarily restrictive. Nevertheless, as a first step in understanding the idea would need to be demonstrated experimentally. My first experiment was to create a vortex and then introduce the charged particles to test out the centrifuging idea of charge separation in the expectation of generating an electrical discharge.

With no ball lightning specialists within the Physics Department I decided to seek advice about how to carry out initial experiments to generate a vortex. My plan was to first introduce a tracer into the vortex to map the movement of air flow. My next step was to inject charged particles to produce a luminous vortex. A visit to the Mechanical Engineering Department revealed that no one had any vortex air tracer experience although they did carry out wind tunnel experimentation. So I was referred on to Professor Williamson of the Chemical and Process Engineering Department. I went over to the department and as it happened Professor Williamson was away overseas. I walked past the door to the office of Dr. John Abrahamson and before long I was talking about my ball lightning proposal. I was lucky. John Abrahamson's special field was directly related to the ideas I was interested in. He worked in the field of cyclones which are engineering devices used to separate particles from a vortex. The geometry of the cyclone is so designed that a vortex is formed inside a metal chamber and separates particles by a centrifugal action. His field was right in line with what I wanted to do experimentally with vortices. Oddly enough, I had conceived of this vortex centrifugal idea only a few weeks before I had met John. I had even put the ideas into a draft paper that postulated a charged vortex capacitor as I have mentioned earlier.

We engaged in a long afternoon discussion of the electrical vortex idea. I also showed him my draft paper on the capacitor vortex notion which he later went over. Instead of merely informing me of how I might go about doing my proposed experiment he agreed to take me on to do the research. I was now able to carry out my Physics Master's thesis work on ball lightning within the Chemical and Process Engineering Department but under the official umbrella of the Physics and Astronomy Department.

The Master's thesis was an opportunity to thoroughly test my idea of a luminous vortex by experimental investigation. This work is documented in more detail in Coleman (1990). I designed and built, with the assistance of my supervisor, and the Chemical and Process technicians, David Brown, Neville Foot, Bob Gordon, Ian Murray, and Ron Boyce, a 0.9 meter diameter vortex chamber to simulate atmospheric vortices, like the dust devil. The cylindrical chamber was made from polycarbonate and wood vortex chamber and had a series of adjustable vanes at the inlet base which imparted vorticity to the air mass as it was sucked into the chamber.

Vorticity is a kind of measure of how much rotation there is in localized moving fluid, which in the case of an atmospheric vortex, is mostly an air mass.

Figure 2 : The 0.9 m diameter vortex generator used in the High Voltage Laboratory University of Canterbury to generate vortex breakdown.

The "spherical vortex" is vortex breakdown

The vortex chamber was first installed in the old grinding room of the Chemical and Process Engineering Department. It was here that I was able to produce the elusive "spherical vortex" which I believed was the essential skeletal form of ball lightning. The sought after vortex "bubble", produced in the vortex generator, was called "vortex-breakdown". It was not by any stretch of the imagination, a perfect sphere. It was globular, oval etc. but not spherical. In fact, photographs of vortex-breakdown in the literature are not spherical either. Later I will suggest a reason why I think that a sometimes a flame within vortex-breakdown might be close to spherical despite the average non-combusting vortex-breakdown departing from this geometry.

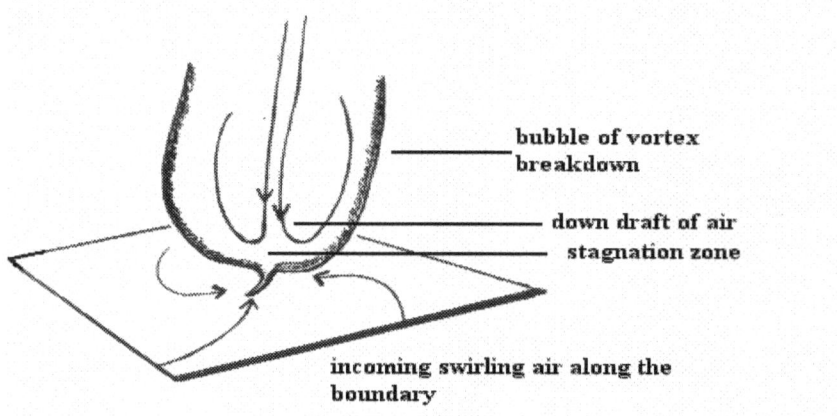

Figure 3: Diagram of showing the main air flow direction in experimental vortex-breakdown

Vortex-breakdown is a lateral expansion of cylindrical core resulting in a pronounced bubble-shaped recirculation region. I will discuss vortex-breakdown in more detail in the next chapter. I found the vortex break down region was invisible to the eye unless air flow tracers like fog dew or neutrally-buoyant-helium-filled bubbles were introduced into the air flow. In the latter method, the eye can track these bubbles as they move along the air flow paths where they are trapped in the stagnation regions of low air speeds for a few seconds. In the case of the fog dew the whole envelope of the vortex-breakdown region was seen and looked very much like photographs of some dust devils I have encountered. Vortex-breakdown appeared for the first time like a mirage out of the mist. It took on a large bubble-like form using the fog tracer and the torch backlighting. The unusual behavior and structure convinced me that vortex-breakdown was the skeleton form of ball lightning without the luminosity. I photographed vortex-breakdown, took notes on its behavior, and read the literature on the subject. I was hoping to find in the vortex-breakdown papers, or by experiment, a specialized property that might unravel the luminosity problem, which was always uppermost in my mind. Even though I investigated an electrical mechanism I was open to the idea that a solution to the "luminous vortex problem" might not necessarily be down this path. But for the time being my working hypothesis was to consider an electrical mechanism. I suspected that these vortex-electrical experiments would either reinforce the electrical approach or suggest an alternative route. I was confident that whatever the outcome of these tests I would eventually discover the solution to the ball lightning problem.

Figure 4: Neutrally-buoyant helium-filled soap bubble in stagnant zone of vortex-breakdown. (Coleman, 1990)

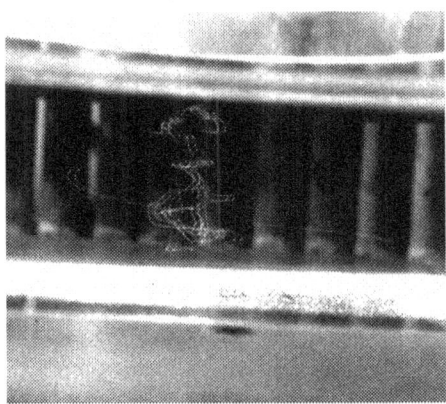

Figure 5: Neutrally-buoyant helium-filled soap bubble spiraling down vortex-breakdown and into the stagnation zone. (Coleman, 1990)

Searching for a luminous vortex.

The first experimental trials were concerned with testing the idea that a vortex had the capability of generating an electrical spark by tribo-charging (i.e. by friction) dust and sand particles, as it lifted off the chamber floor and swirled into the "breakdown" region. The glass window was the central part of the vortex chamber floor. One could look directly upwards to see the bottom of the rotating vortex.

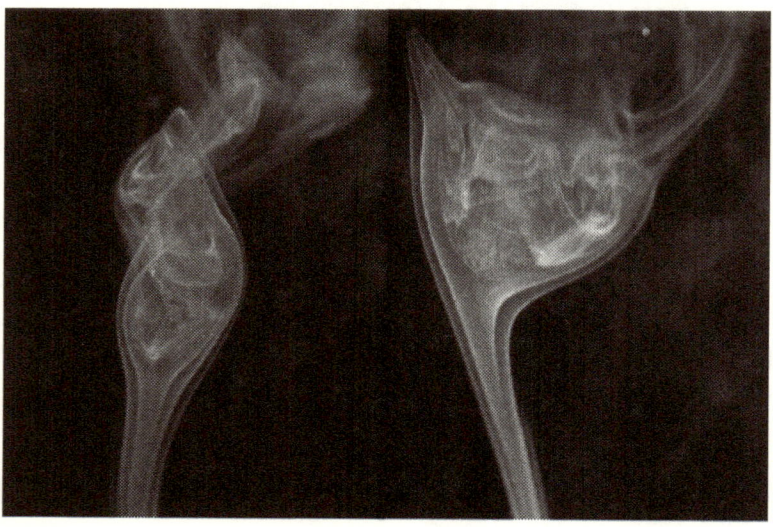

Figure 6: Two examples of experimental vortex-breakdown. One is an oval and the other is a cat-like structure with two-tails and may account for two tailed ball lightning reports. (Coleman, 1990)

I used fine river silt from the Edwards River located in the Arthur's Pass region. In addition, I injected: fly ash, pollen powder, graphite powder, volcanic dust (from the Port Hills), as well as tiny spheres of aluminum oxide powder from the Chemical and Process Engineering Department as dust material into the vortex. The tribo-charging process was supposed to separate charge using the concentric sheaths idea, although I was open to any electrical charging process to make its presence known in the vortex chamber.

An electrometer was borrowed from the Physics Department to detect the effects of this charging process. I discovered that when the electrometer probe was fixed in the vicinity of the vortex-breakdown I recorded a jump in the voltage reading of the probe. This effect coincided with the lifting of sand or dust off the central glass section as it was swept into the vortex. Although this electrical effect was encouraging, I had hoped to see tiny electrical sparks or corona in the vortex as the sand swirled up into the vortex and then into the ducting which led outside to a large tree. This ground charging effect has been observed in dust devils though charge densities are moderate ($\sim 10^6$ e/cm^3) and there is a dipole distribution of charge in a vertical line.

I carried out most of the trials at night and some by day, using a black plastic shroud to produce blackout conditions. I did not see any light generated by an electrical discharge whatsoever in the vortex. I then introduced dust of opposite charge into center of the chamber through the spaces between the metallic vanes. Still there were no luminous effects to be seen.

Charged aerosols

I found a paper by Vonnegut (1966) in which he created clouds of highly charged oil droplets which actually produced luminous electrical discharges. He said that titanium tetrachloride would also be a suitable aerosol to charge. His work inspired me to construct special high voltage electrostatic cylindrical charging tubes, which I suspended in the vortex chamber. Even with this arrangement I could not produce charge densities high enough to generate an electrical discharge within the vortex. I was convinced that my charging precipitation tubes were not generating sufficiently high charge densities to get a spark from the aerosol clouds.

I needed a device that would give me an even higher charge density. A paper by Barretto (1969) seemed to provide a method to increase the charge densities required to generate a discharge. The authors built two compact high voltage cylindrical charging "guns" that ejected highly charged aerosol jets. At the intersection of the two jets, Barretto was able to photograph the existence of electrical glow discharge regions. I decided to try and repeat the experiment using their design so I had two custom-built ionizers made in the Physics Department workshop. A charge density device invented by Obolensky was also made by the Chemical and Process Engineering technicians. This "ion detector" sucked charged aerosol clouds through a filter that was able to measure the volumetric charge density from my aerosol guns. Although I was able to increase the charge densities up to 0.0052 Cm^{-3} (one order of magnitude less than Barreto), it was not sufficient to generate an electrical discharge. With these null results I started questioning whether it was realistic to expect a small vortex to generate a spherical electrical discharge producing the luminous form I suspected to be taking place in naturally-occurring ball lightning.

I explored another avenue. I found a precedent for producing light emission in a vortical flow in the laboratory. Lavan and Fejer (1965) produced light emission in supersonic swirling flow from water droplets in the flow impinging on the cylinder wall of their swirl flow apparatus. There existed at least two significant problems with this sort of experiment for it to be useful to the ball lightning puzzle. One problem is that the supersonic flow condition is too stringent. The luminosity should be able to take place in vortices of much lower air speeds. There is also the problem that the electrical charging effect was a direct result of water droplets affecting a solid section of the vortex generator. In an atmospheric vortex there is no vortex chamber, and therefore, no such extended solid surface area for particles like water droplets to impinge upon and generate charge separation. This luminescence was simply an artifact of the swirl apparatus, and so the luminosity production could not be extrapolated to a naturally occurring vortex. My conclusion was that Lavan and Fejer's study was not relevant to the ball lightning problem.

The vortex generator experiments left me with the conclusion that the original hypothesis of luminosity by electrical discharges in a small dust devil-like vortex was not the answer. There must be some other mechanism to produce luminosity other than by purely electrical means.

I needed another direction to move in. The thesis investigation demonstrated that the electrical mechanism idea was flawed yet I was convinced there was a direct connection between ball lightning and vortex-breakdown. On the positive side the properties of the vortex-breakdown on its own were sufficient to explain several observed properties of ball lightning. The thesis was the first time that ball lightning had been directly linked with vortex-breakdown as the primary structure and UFOs were also partially explained in the thesis. Several qualitative properties of ball lightning could be accounted for qualitatively by invoking explanations involving vortex shape and motion as well as assuming a mechanism operated to keep the vortex glowing. In my *Ball lightning* thesis I devoted a whole section to describing and explaining the characteristics of several specific UFO sightings. I proposed there that a certain type of UFO exists having distinctive properties of shape and motion described by luminous vortex-breakdown and similar to some ball lightning field events. I was still without a definite solution to the luminosity problem, until I revisited a book by Thomas Gold.

GOLD'S BOOK

Thomas Gold has made contributions to many fields of science. He was co-author with Fred Hoyle on the steady state theory of the universe and has been involved in neutron star (pulsar) research. In his book, "*Power from the Earth*", he cited several luminous effects associated with earthquakes. The phenomenon has been termed "earthquake lights". His intention in writing the book was to advance the hypothesis that there is a vast reserve of natural gas locked deep within the Earth. This reserve was thought to be sufficient to provide a large amount of energy for the Earth's population (Gold, 1987). Though more and more natural gas fields are discovered the idea remains highly controversial. If his thesis is correct, it would provide a greater source of gas for combustion theories of ball lightning. It means that there may be a greater abundance of natural gas, than previously supposed, to act as a fuel for ball lightning. It should be noted with caution that Gold's hypothesis has not been widely accepted among geologists though his hypothesis appears to be supported by some documented evidence. I took a special interest in his book since it might be of help with the ball lightning puzzle.

Gold's explanation of "earthquake lights" is that they are flames hovering over the ground sourced by natural gas escaping from fissures in rocks near fault lines. It was only as an aside that he mentioned this and there was no connection to ball lightning stated in the book.

It was one of Gold's quotations of a particular earthquake light observation that suggested to me a crucial insight into the ball lightning puzzle. Wallace and Teng (1980) cited a Chinese scientist, Chu Chieh Cho, of the Provincial Seismological Bureau who reported on unusual fireballs at the Wanchia commune of Chungching County in China on the night of the 21st of July 1976. These fireballs reportedly moved vertically upwards and burnt a hole in a roof of a house. Around one thousand fireballs were seen in the area and were mostly nocturnal sightings. Quite unusual balls of smoke were also seen during the day. It was inferred by the authors that these smoke balls could well have had the same physical nature as the fireballs that took place at night.

The Chinese fireball report appeared to describe something which might have some bearing on the ball lightning problem. This event may have been just the typical chemical combustion fireball and so I was careful not to equate this type of "fireball" reference to ball lightning. Obviously not all fireballs will be ball lightning. For instance, meteors are termed "fireballs" and they are definitely not ball lightning, just as small fireballs a few millimeters in diameter seen in microgravity experiments (Buckmaster et al 1990) are irrelevant to the ball lightning problem. The Chungching fireball could still be the usual type of fireball originating from the rapid combustion or explosion of a volatile liquid or gas. However the reference to small smoke balls was highly unusual and would appear to be inconsistent with the rapid combustion fireball idea. These "smoke balls" might be smoke entrained in a vortex-breakdown recirculation region of a vortex hovering over some location. The description is similar to black ball lightning which I will describe in a later chapter.

Lightning breakthrough

Despite Gold's Chinese fireball reference, and other written accounts of luminosity in earthquakes, I was still without a viable mechanism to explain the luminosity puzzle of vortex-breakdown in relation to ball lightning. In mid 1992, in a sudden flash of insight, I came to a solution while I was browsing through an engineering book on swirl burners, called "*Swirl Flows*" by Gupta et al (1984). Swirl burners are industrial burners designed to use vortex-breakdown to burn a fuel more efficiently. For the first time, it became clear to me that swirl gas burners could exist naturally in the atmosphere. A vortex such as a whirlwind could burn a gas provided the vortex was undergoing vortex-breakdown. I recognized immediately that this was a new scientific idea. Furthermore because it was a previously unrecognized phenomenon perhaps it was also the solution to ball lightning sought for by the physicists.

I had read about atmospheric vortices through the work of Vonnegut and other investigators and I realized that I was on to an original and exciting line of thinking. I started to imagine how ball lightning could be produced by a single lightning stroke. The lightning would act as an agent to simultaneously produce a small vortex and also ignite any methane at ground level. Sightings of larger diameter ball lightning would be interpreted as much bigger vortices containing much bigger balls of fire. Once the vortex is created there is an independent entity having the properties of a vortex, such as darting movements with or against the wind, vortex splitting (which I will mention later) and so on. Lightning, under this theory, would possess only a minor role, since lightning would not be needed, on every occasion, to produce ball lightning. The vortex combustion idea was strikingly simple since only three elements would be needed to create ball lightning: the fuel gas, vortex-breakdown, some ignition source (such as an electrostatic spark) to ignite the fuel gas, and an oxidizer in the form of oxygen in air.

One obvious prediction follows from the vortex burner hypothesis. Ball lightning phenomena should be observed more frequently where there is a high probability that all of the above conditions coincide. I predict that an ideal site for ball lightning creation would be during specific volcanic eruptions where vortices, nature gas and a source of ignition would be present. It then becomes immediately clear why large number of ball lightning fireballs (not hot incendiary rocks) have been seen during some types of volcanic eruptions (Singer, 1971). My next step was to quickly publicize the theory and claim priority in the scientific realm-as scientists do.

4
The Vortex Burner Hypothesis

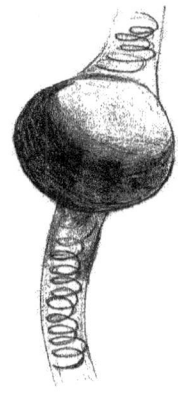

"...a small whirlwind, about 5 feet in diameter and sometimes 100 feet high, formed over a corn-field where it tore up the stalks by the roots and carried them with sand and other loose materials high into the air. The body of the whirling mass was of vaporous formation and perfectly black, the center apparently illuminated by fire and emitting a strange 'sulphurous fire' that could be distinguished a distance of about 300 yards, burning and sickening all..."

PUBLISHING THE THEORY

An opportunity arose to publish my theory when I came across an interesting letter by the scientist, Dr Vonnegut. He wrote to *Weather*, a publication of the Royal Meteorological Society's of the UK. In his letter he discussed an unusual fireball embedded on the axis of a tornado. By composing an appropriate response to Vonnegut's letter I was able to launch my ball lightning hypothesis for the first time in the open scientific literature. *Weather* was an appropriate forum to air the hypothesis since it had a good track record of publishing items on ball lightning.

Vonnegut's letter was a reply to a letter by Brown (1990) who reported on a rather unusual tornado near Dorset that contained a large fireball on its main column. A tornado seen by a farmer of East Stour moved and twisted with a ball of fire on its funnel. Typical tornado damage was observed such as the uprooting of trees.

Presters

Vonnegut's correspondence to *Weather* contained a few more references to luminous activity in tornadoes. He used a generic name for these tornadoes which he called "presters". He did not coin this term because the word "prester" was already in use. It was defined in an old dictionary called the "Latin English Lexicon" of 1851 as,

"a fiery whirlwind that descends in the form of a pillar of fire."

I consulted the New Shorter Oxford English Dictionary of 1993 and found out that the word "prester" had two meanings. The first is *"a mythical serpent, the bite of which caused death by swelling"*. The second meaning was *"a scorching whirlwind"* which seemed to agree more or less with Vonnegut's definition. This dictionary referred to an older English-Latin etymology for the second meaning. Why does the word "prester" not appear as a common entry in modern English dictionaries? Is there are lesser occurrence of these fiery whirlwinds then, as compared with now? I do not think so. It could be more likely that these fiery whirlwinds or presters, have, in modern times, been simply re-identified as UFOs.

The fiery whirlwind or prester was obviously known to early authors on the subject of tornadoes. Vonnegut cited Ludlum (1976) who wrote a book entitled *"Early American Tornadoes 1586-1870"*. In a typical quotation from this book it was written, *"Some say it resembled a whirlwind of fire and smoke"* (pg 68).

Vonnegut speculated that the Stour Tornado, that contained the fireball, was a prester, and that it's light emission was caused through some electrical mechanism (Vonnegut, 1991). Vonnegut had not deviated from his thinking in his *"Electrical theory of tornadoes"* paper (Vonnegut, 1960). With the publication of his correspondence I had a small window of opportunity to provide an alternative explanation for the luminosity in tornadoes and so present my ball lightning hypothesis to a wide audience.

As an aside, fire whirls are different to these presters or fireball vortices I have proposed in my theory for ball lightning. The firewhirl is a long vortex generated by the rapid combustion of a fuel, usually at ground level, which then creates an intense updraft. They do not a direct result of burning a fuel in the combustion chamber of vortex-breakdown.

Theory published in *Weather*

A reply to Vonnegut's letter could be succinct. I appreciated that scientific communications did not need to be long to be important. I was inspired by a famous piece of scientific correspondence to *Nature*. In 1926 Ulenbeck and Gouldsmit proposed that sub-atomic particles have spin. This proposal immediately explained several puzzles in quantum mechanics leading many to consider this brief letter as one of the most important seminal contributions to the field of quantum mechanics. The letter was published in *Nature* with the backing comments by the well known twentieth century pioneer in quantum mechanics, Niel's Bohr.

I sent off a letter on the 25th August 1992. A fax stating acceptance, was sent back to me by the sub-editor of Weather, Mary Spence, on the 8th October. The letter was accepted for publication and finally published in the January 1993 issue, Volume **4,** No.1, page 30. The original letter:

An explanation of ball lightning?

"Vonnegut (1991) identified the Dorset tornado of December 1989 (Brown 1990) as a kind of "prester" whose luminous activity was possibly of an electrical nature. It was said to have resembled a "twisting ball of fire" to the farmer who observed the phenomenon. An alternative explanation of this luminosity is the combustion of an atmospheric fuel such as natural gas, within the tornado. However any flame would be quickly blown out by the high air-speeds. The only conceivable way combustion could occur is when the tornado undergoes axisymmetric vortex-breakdown where there is a great reduction in air velocity.

Such a phenomenon does, on occasion, take place. Particularly clear observations of vortex-breakdown of the Minneapolis tornado of 18 July 1986 were reported by Pauley and Snow (1988). Once this air speed reduction occurs, then it becomes possible that the flame speed can match the air speed. In fact, this matching will define the flame boundary. In addition, the vortex core expands laterally and a bubble-like recirculation region is established. The "ball of fire" observation might then be explained on this basis.

Actually, the observation of the Dorset tornado appears to bear a close resemblance to an event reported by Campbell (1982) near Crail in Scotland in 1968. There, a whirlwind was observed in the middle of Roome beach in close association with a ball lightning object which had a 200 mm diameter and was 0.5 m off the ground. This sort of observation indicates that ball lightning might well be explained as a vortex acting as a burner of combustible fuel such as natural gas or other."

Peter F. Coleman, Lincoln University, New Zealand.

References.
Brown, R., (1990), Tornado in North Dorset: 21 December 1989, Weather, 45, p.240.
Campbell, S., (1982) Ball Lightning at Crail-1968, Weather, 37, pp.75-78.
Pauley, R.T. and Snow, J.T. (1988), On the kinematics and dynamics of the 18 July 1986 Minneapolis tornado, Mon. Wea. Rev., 116, pp. 2731-2736.
Vonnegut, B., (1991) Presters, Weather, 46, pp. 360-361.

AXISYMMETRIC VORTEX-BREAKDOWN

Since vortex-breakdown is a key aspect of my fireball hypothesis I will describe it in more detail. As I suggested in the letter to *Weather*, the presence of vortex-breakdown makes it possible for a large tornado vortex to act as a flame carrier because of the much reduced speeds associated with vortex-breakdown. In my opinion this reasoning could not have been proposed before about 1957 simply because vortex-breakdown was not "discovered" until then. There have been authors, like Maxworthy (1982), who thought that vortex-breakdown was reported in much earlier experiments with water vortices, like that of Wilke in 1785. Dessin de Michaud also of the 18th Century, sketched a waterspout touching down in the Mediterranean with what obviously appears a round section, two-thirds the way up the funnel. This looks very like vortex breakdown. However none of these observers understood vortex-breakdown as scientists do today.

Discovery of vortex-breakdown

Peckham and Atkinson (1957) are now recognized in the field as the first researchers to discover vortex-breakdown as we now know it. They discovered the vortex phenomenon from the flow separation on the leading edge of wings set at high incidence angles. It is interesting that several fireballs have been seen on wing tips, including the so-called "foo-fighters" of World War 2. Since there are two wing tip trailing vortices there is the possibility of a maximum of two fireballs one on either side of the plane.

Since Peckham and Atkinson's work there have been numerous papers written on the subject of vortex-breakdown-not only on the aerodynamics of aircraft. Vortex-breakdown has been applied to such widely varying topics as tornadoes and swirl burners used in industry. Vortex-breakdown is a relatively new field of investigation and such work would not have been available to early workers in the field of ball lightning.

Theories of vortex-breakdown

Vortex-breakdown is now a familiar phenomenon in the scientific literature though still not well known in the popular literature. Vortex-breakdown still has no definitive theoretical framework but the big advantage is that it can be created with relative ease in the laboratory. A theory of vortex-breakdown is likely to be found in an appropriate fluid dynamic theory. On the other hand ball lightning's nature is a complete unknown to the wider scientific community and some have said it has yet to be convincingly created in the laboratory.

Hall's (1972) widely cited older review is still pertinent to the subject of vortex-breakdown, though there has been later research which claims a better understanding of the phenomenon. Hall considered the study of vortex-breakdown to be one of fundamental significance to both science and technology. He said that the problem has a number of different explanations but that no single theory has been accepted. Hall pointed out that the phenomenon is just one type of discrete steady-state perturbation of a rotating stratified fluid.

Hall divided explanations of vortex-breakdown into three categories. The first was the idea that vortex-breakdown is analogous to the separation of a boundary layer. The second was that vortex-breakdown is a type of hydrodynamic instability. The third approach suggested that vortex-breakdown is a critical state phenomenon likened to a hydraulic jump in which there is a change from supercritical flow to sub-critical flow. Such categories belong to the scientific field of fluid dynamics. The explanations are technical and I will not dwell on these theories any further but refer the reader to papers like Hall (1972) if they wish to begin delving further into the merits and defects of each type of theory. It should be noted that vortex-breakdown is not symmetrical about the rotation axis but has a complicated velocity flow field, not at all like Hill's vortex which is a well defined theoretical solution to the Navier Stoke's equations for fluid flow.

Types of vortex-breakdown

There are essentially two types of vortex-breakdown. One is called "axisymmetric", while the other kind is "spiral breakdown" though there are more detailed classifications. Some workers have recognized three main classes of vortex-breakdown, axisymmetric, spiral and double helix vortex-breakdown. The axisymmetric form of vortex-breakdown is the preferred candidate for my ball lightning hypothesis because it has both a globular shape and a recirculation region with the greatly reduced air speeds needed to allow combustion of the fuel gas. The reduced air speeds are required to prevent the "blow out" of the flame.

The three basic forms of vortex-breakdown have been extended by later research. Faler and Leibovitch (1978) reported seven modes of vortex-breakdown in water vortices, which were designated by numbers running from 0 to 6. In this classification system axisymmetric vortex-breakdown has two types which have been labelled "type 0" and "type 1". These seven modes were increased to eight. Sarpakaya (1995) reported on another unusual conical mode which they found in turbulent vortex-breakdown. Likewise, Khoo et al (1995) a few months earlier discovered a conical envelope form of vortex-breakdown in high swirl conditions. This conical mode may explain certain ball lightning and UFO shapes that look like inverted ice cream cones. A fine example of this may be the Canaries UFO which I will discuss in Part II.

Faler and Leibovitch's studies showed the existence of transformations between these two basic types which they said were dependant on two dimensionless numbers. The first is the "swirl" or "swirl number" which is defined as the ratio of axial flux of swirl momentum to the axial flux of axial momentum multiplied by an appropriate radius parameter. The second number is the Reynold's number. A technical definition of the Reynold's number is that it is a ratio containing a radius parameter, (eg. diameter of the vortex chamber) multiplied by the air flow speed and divided by the kinematic viscosity of air. It has been found experimentally, that for a fixed Reynold's number, which, in practical terms, means a steady rate of air flow, as the swirl (or swirl angle) is increased, the spiral form eventually transforms into the axisymmetric form. The change is usually fairly abrupt.

Physical description of vortex-breakdown

The physical description of the little understood axisymmetric bubble vortex-breakdown is that of a vortex core which, at some point, suddenly swells into a bulbous shape. As the flow spirals up around the sides of the bubble it reverses its flow and then feeds back into the bubble. This pronounced reversal also results in a slowing down of fluid flow as it moves in two main recirculation loops analogous to convection currents in a cup of coffee.

Within the recirculation region there exist so-called "stagnation points" characterized by very low air speeds. One major stagnation zone is located vertically above the point where the cylindrical core starts to expand laterally to form the bubble. These stagnation points have been located in experimental vortex-breakdown regions. I detected these zones using soap bubbles in air vortices. I observed them (Coleman, 1990) by using neutrally-buoyant and helium-filled soap bubbles which were delivered into the vortex and became trapped in a sometimes tight acrobatic path with the helium-filled soap bubble bobbing in and around this zone for a few seconds. Using strobe photography and later analysis, I was able to show that the velocities in this region were much lower than the main vortex flow which was typically around 1 ms-[1].

Detailed studies of the internal structure of vortex-breakdown

Since my unified theory incorporates vortex-breakdown the more knowledge that can be gained in this area the better and would lead to a better understanding of ball lightning. Most of the researchers in vortex-breakdown would be unaware that their work has a direct bearing on elucidating the age-old problem of fireballs.

Much of the work on vortex breakdown has been under controlled conditions using well defined containers such as cylinders with a rotating base and liquids with a much higher viscosity than air. For example, Faler and Leibovitch (1978) detected stagnation zones in their small water vortices with specialized laser equipment using the Döppler effect which can measure fluid flow speed. This measurement technique is called "non-intrusive" because it does not interfere with the fluid flow. It is considered important that such non-intrusive methods be used because vortex-breakdown is known to be super-sensitive to slight changes in flow. They used experiments employing a technique called laser Döppler anenometry (LDA) measured velocities of the fluid and found the inner cell does not have reversed flow along the axis. The inner cell is bound between two stagnation points along the axis. The existence of the inner cell has not yet been confirmed either in natural vortices by observation or direct measurement, but going on the track record of features observed in laboratory flows that have subsequently been found in natural flows it will eventually be detected. Leibovitch (1978) reported that the inner cell has been found in the work of Sarpakaya (1971a) and the short note by Bellamy-Knights (1976).

Sotiropoulos and Ventikos (2001) generated numerical vortex breakdown bubble of liquid rotating in a cylinder and compared their results to experimental studies using electrostatic powder particles. The authors claim that such confined flow studies have become a standard way of working out and understanding the basic internal dynamics of the bubble form of vortex breakdown. This claim was made by extrapolation from studies with confined flow in a closed cylinder. There was no supporting argument as to why these discovered features of their bubble vortex breakdown should appear in other contexts. A fundamental explanation of the underlying physics has not been revealed and so the extrapolation to other areas of study appears to be based on more of an intuition. It could well be the case that this flow structure is discovered in the bubble form of vortex breakdown in other contexts. The authors make no mention of the connection with the field of meteorology since their primary interest is in the engineering context. Their numerical computations used the usual Navier Stokes equations for fluid flow and confirmed the existence of toroidal flow in a direction around the axis as in previous studies. This toroidal flow may help the following explain ball lighting shapes; the torus, the wheels within wheels UFO observations, and disc-like structures embedded in the main sphere. But the combustion of fuel needs to originate from stagnation zones. Their final three dimensional topological diagram obtained by a Cray T-90 computer and their graphics package shows the features of the spiraling axial down flow and toroidal loops.

Chaotic paths in ball lightning and some UFOs?

What is interesting is that Sotiropoulos and Ventikos, following a suggestion by Holmes (1984), demonstrated that their vortex breakdown bubble had chaotic paths for particles in the interior of the bubble where there is a generally emptying and filling of the bubble. This feature is unable to be modeled using axisymmetric numerical studies as Leibovitch had already pointed out. Further experimental studies by Sotirpoulos *et al* (2002) confirm this chaotic behavior. If such chaotic behavior is valid not only for their glycerine and water (3:1 by volume) vortices then such chaotic behavior may eventually be revealed in vortex fireballs. There are a number of other features observed in this type of study that could also exist in real vortices.

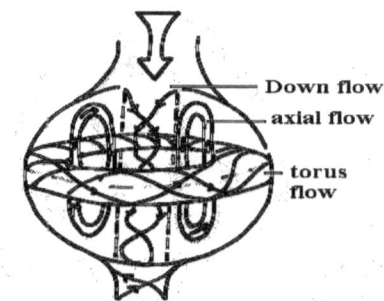

Tori inside vortex-breakdown

Figure 7: Toroidal flow in vortex breakdown

A CONCEPTUAL MODEL OF THE VORTEX BURNER THEORY

I do not intend to present a quantitative model of the fireball theory. A realistic treatment would come from other researchers and will take some time to develop. Research from the two fields of flame combustion, vortices and vortex breakdown would need to be applied. Each of these fields is loaded with sub-problems and much work needs to be done. I will, instead, give a broad brush model setting the stage for later work.

It is well known that twisters can lift material into the air and retain it for long periods of time. Thus the general idea that the vortex acts as a kind of crucible for combustion is better able to deal with many other types of vortex fireballs. Two types of vortex fireball could be distinguished- the fuel gas fireball and the solid combustible fireball. Any solid fuel remaining in a vortex could ignite a fuel gas like natural gas. This might explain how a vortex would be seen to flare up when there is a sudden injection of new fuel-rich gas.

Solid combustible fireball model

Trees, branches and other solid combustibles could be lifted into the vortex and combusted. This material could be retained with the vortex-breakdown because of the lift force of the vortex balancing the gravity force. There is also the balancing of the drag force and the centrifugal force. If there is an abundance of combustible material, like wood, that could be drawn into the vortex and retained. Lightning could then strike the combustible in the vortex and set it on fire. The solid fuel fire inside the vortex existing as glowing embers, could suddenly come back to life with new combustible material. This would effectively extend the lifetime of the luminous fireball.

What other material could be burnt by the vortex? Wood and natural gas are common but there could be finely suspended combustible particulate material like coal dust or pollen. Under this model if there is some object already burning in the vortex anything combustible will be drawn into the vortex and burnt. Very fine iron dust could also be a potential fuel. This would be a much brighter flame than with straight wood combustion. This returns us full circle to my earlier ideas of particulate material brought into a vortex and centrifuged outwards to form particle sheaths.

This solid combustible vortex fireball model might explain cases where ball lightning is seen in areas devoid of natural gas emission. So the overall theory is not necessarily restricted to gas rich areas. But whatever fuel is burnt invariably the site of initial combustion must come from within the stagnation region of vortex-breakdown otherwise the flame would be quickly blown out.

Solid fuel vortex burners could show some of the familiar features like a fire burning in a fireplace but within a swirling air environment. There will be ash connected with this type of burner. Some of the solid fuel may even fuse together. Industrial solid fuel swirl burners would provide valuable comparisons with the natural counterpart.

Likely flammable gases involved

It is important to realize that the viability of the theory of combustion inside a vortex does not rest on any one fuel. A few critics have attempted to dismiss the theory by saying that no natural gas exists at this or that location. They might say that the flammable gas is not emitted at this or that location, when in reality the possibility does exist. Many naturally occurring gases, like carbon dioxide, are not flammable. The two most likely flammable gases for the vortex burner theory would be natural gas and hydrogen sulfide gas. Note that hydrogen sulfide can also be present in natural gas.

Before I concentrate on natural gas I will briefly describe hydrogen sulfide, since it is expected to account for a proportion of ball lightning and UFO reports. Hydrogen sulfide has the odor of rotten eggs which has been reported in both ball lightning and UFO cases. Hydrogen sulfide is heavier than air and is highly flammable (the flammable limits are 4-46 %). The gas burns with a blue flame and auto-ignites at 260 degrees Celsius. The UFO investigator, A.F. Rullán, wrote a report called *"Odors from UFOs"*. He concluded that if the chemical sources of the odorant were gases, then hydrogen sulfide was a likely candidate among the five other gases listed.

Hydrogen sulfide from volcanoes

Many volcanoes emit hydrogen sulfide gas along with greater proportions of water vapor and carbon dioxide. The investigator, E. Bach even wrote a book suggesting that UFOs actually came from volcanoes. Sightings of UFOs from ordinary active volcanoes could be accounted for by the theory described here. The high temperatures in the vicinity of the volcanoes could easily ignite hydrogen sulfide inside vortex breakdown. The sightings and photographs of fireball objects around active volcanoes like the Mexican volcano Popocatepetl ("Popo"), in the early 1990s, and later, are thus explained. Whirlwinds around volcanoes (even up to a kilometer away) have been reported in the literature by Thorarinsson and Vonnegut (1964) and others. They observed vortices downwind of the volcano.

The hot convective updraft of hot gases, one of the conditions required for a dust devil-like vortex is present. Wind shearing across this could produce angular momentum to give rise to a twister. A stream of hot gases is characteristic of active or erupting volcanoes. Therefore it is no coincidence that the UFOs seen and photographed over Popocatepetl commenced when the volcano became active around 1991 after a number of years of inactivity. Some photographs on the internet reveal that these UFO sightings seen at Popo were also downwind.

Fuel gas fireball model

An idea of the fuel gas burner hypothesis can gained by imagining the movement of fuel gas like natural gas up the upstream vortex core and then into the burning region of vortex-breakdown. What we have is an atmospheric vortex with no solid tube to conduct the gas to the burner region as in a laboratory burner like a Bunsen burner. However the vortex core basically acts like the barrel of the burner. In metal pipe burners it is easy to produce a predominantly diffusion flame, or pre-mixed flame, by controlling the air to fuel mix before combustion. Possibly in an atmospheric vortex mixing of the fuel gas with air (containing oxygen), could take place prior to entry into the recirculation region. This will no doubt depend on such things as the extent of turbulent conditions in the core which feeds into the vortex-breakdown. Turbulence tends to disturb pure streams of fuel gas and some entrainment of fuel leading to mixing would seem inevitable. Perhaps a diffusion flame may dominate in low speed laminar air flow. Note that natural gas is a mixture of gases and is typically stated as having a natural gas composition such as 85 %

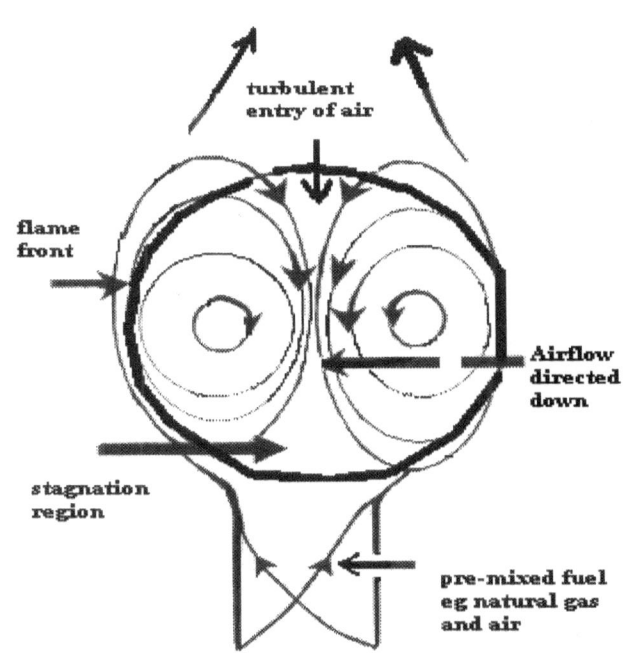

turbulent entry of air

flame front

Airflow directed down

stagnation region

pre-mixed fuel eg natural gas and air

methane, 10% ethane, 3% propane plus 1% other gases. But natural gas compositions can vary greatly. One source in Kentucky has been reported as having 69.7% ethane (Jones, 1985).

There could be an efficiency advantage in burning gas in an atmospheric vortex-breakdown. The entry at the rear of the bubble is quite turbulent and could enhance the mixing of fuel and air to produce a well aerated and efficient flame. The fuel gas concentration can be rich or lean and be in the flammable limits (eg lean limit 5% and the rich limit of 15% for methane). Another advantage is that as the fuel gas moves up into the combustion region it will become pre-heated- a definite advantage for efficiency. Turbulence is also reduced for high swirl flows thus stabilizing the flame.

The flame front is controlled by the aerodynamics of the recirculating air within vortex-breakdown. Some research suggests that there needs to be a matching of the air speed and the flame speed (assuming a pre-mixed flame) such that if the air speed is greater than the combustion wave the flame would be blown out. If the flame speed is greater than the air speed there would be "strike back". This could happen in a natural vortex. The flame suddenly moves into the low air velocity region. A typical flame speed for natural gas and air combustion in a tube is 0.6 ms-[1] (Jones, 1985), though this is dependent on pressure. Speeds within vortex-breakdown need to be of this order of magnitude.

In some cases the flame of ball lightning will be a turbulent diffusion flame but and in others a turbulent pre-mixed flame, especially for fireballs in tornado-sized vortices. Laminar flames take place at low air speeds. Pre-mixed flames based on existing swirl burner studies are noisy and can be described as a hissing sound. This is just what some ball lightning objects have been reported to do. Some authors have put the hissing down to corona but the other alternative is a combustion noise of a turbulent flame or the noise of air in the vortex.

As with many flames the ball lightning spherical flame may contain both diffusion and pre-mixed regions (Gaydon and Wolfhard, 1953) and this is likely to be true for the swirling vortex-breakdown flames as well. In some flames I envisage a smaller pre-mixed flame inside a diffusion flame envelope rather like a candle flame. If the outer diffusional flame envelope was optically-thin an observer should be able to see right into the ball lightning and see the blue colored pre-mixed part of the flame. This kind of internal structure has been reported of ball lightning. But much of this fireball flame analysis needs to be investigated both in the field and the laboratory.

Miscellaneous aspects of gas combustion

The early pioneering work of Gaydon and Wolfhard (1949) found that in their spectroscopic flame studies the blue to violet emission was due to stoichiometric mixtures of methane and air burning and producing OH and CH spectral bands. Stoichiometric proportions are the exact amount of fuel gas and oxidizer required for complete combustion. For example, the stoichiometric proportions for methane and oxygen would be according to the chemical equation, *1 vol Methane + 2 vol oxygen = 1 vol carbon dioxide + 1 vol water* That is, one volume of methane gas is required to combine exactly with two volumes of oxygen. For a methane and air stoichiometric mix the percentage by volume required of methane is about 9.5%. Here there is no excess or deficit of either methane or oxygen. Cases of ball lightning involving the combustion of natural gas in vortex-breakdown are likely to show the same or similar methane emission spectra.

In the stagnation region of vortex-breakdown concentrations of natural gas into the fireball could still produce a flame even when other flame regions of the vortex-breakdown are extinguished. The vortex fireball theory predicts a kind of "pilot light" region inside atmospheric vortices. This effect, to be described later, was confirmed experimentally in a vortex chamber. The impression to an observer might be that the light went out when what really happened was that the spherical flame shrunk to a small region of light in the vicinity of the stagnation zone only to later flare up again. Both ball lightning and UFOs have, in the past, exhibited this type of behavior.

Flame combustion involves complex chemical equations leading to the production of what are considered pollutants such as NOx i.e. nitrogen oxides and other pollution products. Thus the theory predicts the production of these products in connection with vortex fireballs.

Flame transformations

Since the basic flame shape will be altered by changes in the air flow pattern and mixture ratio it is logical to suggest that this mechanism may account for the distortions in the basic shape of ball lightning. Thus it becomes easier to see how, in principle, a spherical flame shape could transform into an oval flame.

Fiber production

Another possibility presents itself when combustion in vortex-breakdown is entertained. If an atmospheric vortex sucks in dust or other particulate matter into the burning furnace, the possibility exists for particulate material to fuse together to form long fibers or threads (Is this an explanation of some "angel hair" phenomenon in UFO cases?) The swirling air flow may even enhance the production of these long fibers, which could also tangle and stay in the recirculation region for a time. The possibility is that the fibers could become electrically-charged if the vortex was created by lightning this might explain ball lightning's apparent electrical attraction to certain objects.

THE FOURTH INTERNATIONAL SYMPOSIUM ON BALL LIGHTNING

Advantages of a vortex-breakdown chemical theory

When I saw an advertisement by the International Committee on Ball Lightning that called for papers for their 4th International Symposium, I began by writing a paper entitled *"A combustion hypothesis to explain ball lightning"*. I had a chance to add more detail to my hypothesis contained in the letter to *Weather*. My paper aimed to put the case for the vortex burner hypothesis clearly in the realm of chemical theories, as discussed by Singer (1971). In the paper I pointed out some of the problems other ball lightning theories have encountered and the advantage my theory has in overcoming the problems connected with chemical theories. Here is an extract from the introduction:

"Many hypotheses have been proposed to answer the question surrounding the nature of ball lightning (BL). Singer (1971) critically reviewed several hypotheses and found that many of them were unable to fully account for the diversity, including observational data, such as: lifetimes, size, shape, texture. and motion. Singer also pointed out the difficulty that chemical theories involving clouds of fuel gases have, in explaining movement against the wind. Clouds of fuel gases would be quickly dissipated. Nuclear theories have many difficulties. Altshuler, House and Hildner (1970), in proposing a nuclear theory based on radioactive isotopes of oxygen and fluorine stated that there were numerous and difficult theoretical objections to overcome if ball lightning has a nuclear source of energy. Jennison (1971) raised a number of objections to the more exotic antimatter hypothesis proposed by Ashby and Whitehead (1971).

Charman (1982) pointed out objections to the major classes of BL theories. Chemical theories were said to have difficulty in explaining why fuel gases should concentrate into a spherical region, and if they did, the glowing region would tend to rise because of buoyancy forces. Hypotheses which postulate the ionization of air molecules have short recombination times, much shorter than the observed lifetimes of BL. In ionization theories it is difficult to see how ions should be preferentially brought into spherical regions rather than be extended along the lightning path. Microwave hypotheses also present difficulties. The major difficulty is that the measured intensities of radiation in the microwave band, under natural conditions, are less than that required by this class of theory.

Singer (1963) singled out one severe constraint on hypotheses involving the transformation of a lightning stroke into ball lightning. The resultant spherical lightning would need to last longer than the actual lightning discharge.

Powell and Finklestein (1970) provided a brief critique of some BL theories. For example, hydromagnetic confinement models, exchange force models of BL., and miniature thundercloud models were said to be inconsistent with the law of continuity of energy and momentum in which there is an upper limit on the energy content of a ball of gas. Radioactive models require large energies to from short-lived isotopes.

In this paper, I will describe an hypothesis which can be demonstrated experimentally, and overcomes some of the usual difficulties attributed to chemical combustion theories. The hypothesis can also be applied to explain a number of BL observations."

In this paper I stated how the vortex-breakdown burner hypothesis was able to overcome the traditional objections that are leveled against the chemical class of theories.

"The main advantage of this combustion hypothesis is that it provides a mechanism to overcome the main objections associated with chemical theories. An atmospheric vortex can sweep up fuel gas, such as methane, over a wide area and concentrate this gas into the stagnant region of vortex-breakdown. Confinement of the fuel gas within vortex-breakdown also prevents dissipation from the prevailing winds."

Practical demonstration of combustion in vortex-breakdown.

I decided to demonstrate combustion within vortex-breakdown using a fireproof vortex chamber although burning a fuel in vortex-breakdown type air flows has been achieved in swirl burners. However, the geometrical arrangement and air flow patterns of these industrial burners are quite different. The very first type of vortex chamber I used was similar in form to other vortex chambers used to model atmospheric vortices. The cylindrical steel pipe chamber was 98 mm in diameter and 2.2 meter high. A small vortex chamber would be ideal because in larger chambers there is a danger of a gas explosion. At the base of the pipe was a detachable steel vane assembly consisting of ten vanes. I built the vane assembly at the Natural Resources Engineering Department at Lincoln University. The baffle assembly consisted of small aluminum tubes. This baffle was placed inside and at the top of the main vortex tube. A baffle is a standard item in vortex generators to promote laminar or smooth air flow. A galvanized iron plate 2 mm thick acted as the base for the vane assembly.

A hose from a standard vacuum cleaner was connected and joined to the top of the steel pipe via a flange joint. The vanes were orientated so that each one was fixed at the desired angle of 30 degrees from the radial direction. The fuel gas was delivered by a plastic hose attached to the nozzle of a butane camping stove. The vortex generator was portable and could be easily set up at any given location.

After switching the vacuum cleaner on, I checked the existence of the vortex-breakdown recirculation region. I inserted a piece of cotton thread between the vanes and into the center of the vortex chamber. When the cotton flicked downwards I knew I had breakdown. This type of method is a standard one to detect reversal of flow and has been used by other workers, like Vu and Gouldin (1982) who used a small piece of wool tuft to find the region of flow reversal and confirmed this with a pitot probe pressure measurements.

My instinct suggested that any piece of cotton thread that I placed through the vanes into the air flow would immediately be sucked straight up the chamber into the vacuum cleaner. When the thread was not at the correct location it was quickly swept up. But at a certain position the thread remained suspended in the air flow for several seconds. This effect is startling to observe. Once I had confirmed the existence of vortex-breakdown I set about introducing butane into the vortex via a small plastic tube. A small but vigorous crackling flame popped into existence well inside the chamber. This flame was fed from butane coming from the plastic tube placed 115 mm away. This small-scale experiment may imitate in some way a fireball in nature which would be fed by one or more natural gas sources on the ground and then sucked up and mixed with air in a flammable fuel-to-air ratio into the burning vortex-breakdown.

The dancing flame

The small flame, slightly less than a centimeter in diameter darted around the floor of the chamber and after a few seconds it extinguished, probably from irregular fluctuations in the air-butane ratio which may have deprived the flame of its fuel. I could bring this flame into existence easily and it convinced me that the burning vortex flame phenomenon was reproducible. If my hypothesis is a correct description of the nature of ball lightning this simple demonstration may well be the first time that ball lightning would have been consciously created in a controlled environment and identified with combustion in vortex-breakdown.

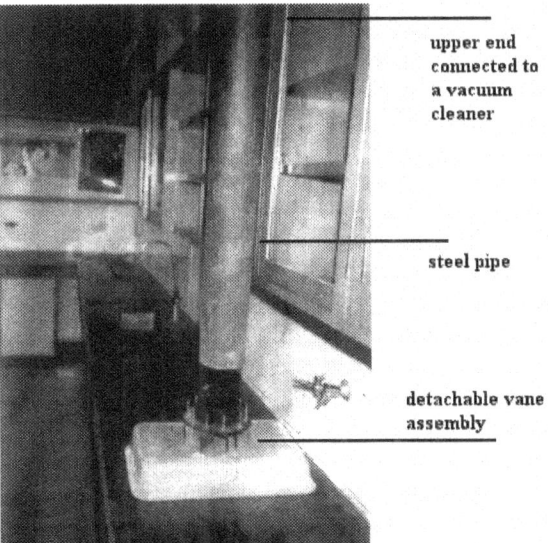

upper end
connected to
a vacuum
cleaner

steel pipe

detachable vane
assembly

Figure 9: The first laboratory creation of a vortex fireball

Barry's yellow-green fireball identified.

Such a vortex combustion phenomenon may have been created in the laboratory before but not identified as such. The yellow-green ball produced by Barry (1967) in his high voltage experiments seems to have satisfied some of conditions for a burning vortex. A high voltage spark was discharged in a box containing a low concentration of propane. A question that remains is this. Did the electrical discharge create a vortex which then manifested itself by whizzing around the chamber burning up the propane? I suggested that this may have happened in Barry's experiment in my paper to the 4th International Symposium on Ball Lightning.

"The experiments of Barry (1967) are interesting in this regard. A yellow-green ball was produced from an electrical discharge into a chamber containing propane with a volume concentration (1.4 % to 1.8%), below the stated lower limit of 2.8% necessary for combustion. Evidently some process must have acted to concentrate the fuel gas so that the fuel-to-air ratio lay within the flammable limit. Vortex-breakdown provides a mechanism to concentrate the fuel gas, and may well account for this laboratory fireball."

Could the engineers who have used swirl burners and vortex-breakdown be said to have created ball lightning? Their chambers were not designed to produce naturally-occurring atmospheric vortices but from an engineering point of view, to act as combustors to burn a liquid or gaseous fuel more efficiently. It comes as no surprise that although such flames are related they are conceptually different because their burners were producing a flame from a nozzle burner. The motivation behind swirl burner studies was quite different to the aim of investigating the ball lightning problem.

Experimental Work in 1998

Since the first edition of my book *Ball Lightning-A scientific mystery solved* there have been new results and photographs of the work carried out in 1998. A much larger vortex chamber was designed to burn a flammable gas at the Chemical and Process Engineering Department within Canterbury University. My original 0.9 meter diameter vortex chamber was re-designed with the polycarbonate chamber replaced by a galvanized iron cylinder. Heat-resistant material was placed in the chamber floor containing a piece of heat resistant glass in the shape of a square. This glass allowed viewing through the bottom of the chamber.

John Abrahamson also suggested that four jets be inserted in the bottom of the floor. These jets were to introduce gas along the boundary layer into the vortex from copper piping. A carbon rod arrangement with a cable to a high voltage source provided a spark to ignite the gas. The idea of igniting a volume of combustible gas is dangerous especially if the gas mixed with air could enter the explosive regime. A solenoid switch-operated valve fed a natural gas (fed mainly methane in the first trials) through the four jets. The spark rod was introduced into the chamber and switched on. Much to my disappointment the spark did not ignite the gas. We altered the vanes to a greater swirl angle. Still nothing happened. Unlike my earlier non-combusting experiments the main blower fan above the chamber was fixed on one speed setting. Other changes would need to be made. I suggested that a heavier mix of gas might work. Perhaps the problem was that the lower density of methane allowed the easy escape of the fuel gas up into the vortex before it could be ignited. The other difficulty could have been that the methane became so mixed with air that it was well outside the fuel-to-air combustion ratio. We produced partial ignition very near a jet entrance but the blue flame was easily blown out. We tried LPG (liquid petroleum gas) but there was still no sustained combustion flame.

Clearly demonstrating combustion inside vortex-breakdown was not going to be as easy as I had imagined. It seemed a contradiction to create a combustion flame in a fast moving air stream. The air speed was high and any flame was extinguished. This is like trying to light a match in the wind.

Figure 10: The combustion vortex chamber used to produce ball lightning (vortex fireballs) in the Combustion Laboratory, Chemical and Process Engineering Department University of Canterbury.

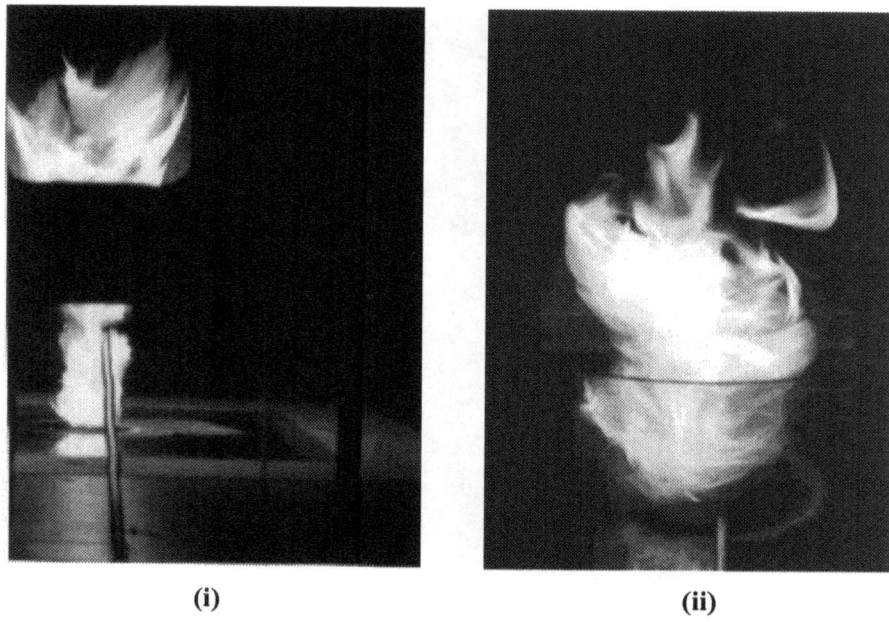

(i)   (ii)

Figure 11: The first pictures of combustion in vortex-breakdown in an atmospheric-like vortex (i) 40 degrees vane angle (ii) 66 degrees.

The four jets idea was not working so we opted to try something else. As if reading my mind about the next step, Neville Foote the senior technician placed a hand-held copper pipe with gas issuing out. Significantly the only location where we could achieve combustion was about five centimeters above the middle of the chamber floor. It was likely that this was the location of the stagnation region, though it was not confirmed when the vortex was in the burning state. But it is in the right position that one would expect the quiescent zone to be located since it was roughly where the helium bubbles were getting trapped in the vortex-breakdown in the original chamber.

At a 40 degree vane angle a conical diverging flame lit up from a small localized stagnation volume. There was a loud combustion sound from the burning flame. When the vane angle from the radial was adjusted to 66 degrees (see photograph) the flame became shorter and more round. I had finally achieved the experimental goal. It was now time to publicize the result. A photograph was reproduced as a color print and published on the front page of the Wellington newspaper, the *Evening Post* in February 1999. The *Christchurch Press* in 1998 featured a short article entitled, *"Burning gas inside tornadoes cause of UFOs"*. This headline spawned two TV interviews. I turned down an interview on the Paul Holmes Show because I thought he might treat the issue superficially. He was more into the "men in black" scenario. Instead I received a sound bite segment on the 6 o'clock TVNZ news of the day. This media publicity was followed up with some local Canterbury TV interviews. The experimental work was published in 1999 in *New Scientist* in an article "*All fired up*" of the 22 May issue.

The scientific communication of Coleman and Abrahamson (1999a) was important since for the very first time the combustion inside a tornado-like vortex in a state of vortex-breakdown was reported in a widely disseminated publication. This respected scientific source was *Eos*, a publication of the American Geological Union (AGU). The archives of the AGU website read;

AGU MEETINGS

HR: 1330h
AN: A32B-01
TI: **Combustion Flame in a Tornado-like Vortex in a State of Vortex-breakdown**
AU: * Coleman, P F
EM: Colemanpf@xtra.co.nz
AF: Chemical and Process Engineering Department, University of Canterbury
     Christchurch, 4 New Zealand
AU: Abrahamson, J
EM:
AF: Chemical and Process Engineering Department, University of Canterbury
     Christchurch, 4 New Zealand
AB: It has been hypothesised (1) that vortices from a few centimeters in size up to a mature tornado may be capable of burning a gaseous fuel, like natural gas. One might expect that any flame would be blown out by the high air speeds in the tornado. However the vortex burner hypothesis proposes that regions of low air speeds within vortex-breakdown, equal to or less than the flame speed should be able to support a combustion flame. We demonstrated this by producing a rotating combustion flame inside small 'tornado' vortices undergoing vortex-breakdown inside an 800 mm diameter chamber an extraction fan to generate an updraft. Sixty vanes imparted swirl to incoming air at the base of the generator. We investigated vane angles of 40 and 66 degrees corresponding to air flow rates of 1.2-0.85 m3/s. LPG/CNG gas feeds were investigated for different flow rates of 1.5-10 NLmin-1 to the vortex via a steel pipe (620 x 6.5mm O.D.)connected to a regulated gas supply. The gas discharge was placed at various positions up to 0.2m above the chamber floor. A spark plug mounted on an insulated rod provided an ignition source. Successful ignition took place at a location along the vertical axis about 50 mm from the chamber floor. This flame

was rotating (with some precession) and associated with the downdraft of the inner cell of the breakdown. The round LPG flame at a vane angle of 66 degrees contrasted with the narrow, upwardly-tapering flame at 40 degrees. The CNG flame less stable, noisier and less well defined than the LPG flame. The existence of a combustion flame within these artificial vortices suggests that such a phenomenon could take place in naturally-occurring vortices. We think that this poorly understood combustion phenomenon may have already been reported in the literature in a variety of meteorological contexts.
(1) Coleman, P.F. 1993, Weather, 48, No.1, 30.
DE: 3399 General or miscellaneous
SC: A
MN: 1999 Spring Meeting

A paper that described the flame experiments in detail (Coleman and Abrahamson, 1999b) was sent to be published in the Proceedings of the 6th International Symposium on ball lightning in 1999 which was held in Belgium and organized by the ICBL especially Professor Dijhuis. They hold such symposia every two years.

An Australian film crew from '*Beyond Horizons*' came over to film the experiment at Canterbury University the following year in 1999. They were tipped off by Mark Stenoff who was a ball lightning researcher and author of the book called "*Ball Lightning- the unsolved problem*" which reviewed the current state of art of ball lightning research. As I said earlier this book follows in the line of other reviews, such as Stanley Singer's monograph. My theory of ball lightning is also cited in the Stenhoff review on page 207. A description of the theory was very brief with no discussion at all of the central idea and how it could explain a number of frequently observed observations. "*Some models suggest that vortex motion of a gas could help to preserve its spherical structure Coleman, 1993, 1997* ".

The film crew stated that in all their travels they had not filmed an experiment creating anything that could be said to be ball lightning and actually conducted at a university facility. For the first time the experiment was shown to millions of viewers around the world on both the Discovery and National Geographic channels.

The importance of the stagnation zone to combustion

The combustion experiments revealed a crucial result. It appears that in order to ignite and sustain a flame inside vortex-breakdown a source of flammable gas in concentrations above the flammable limit must exist in the stagnation zone located upstream where fluid enters the top of the vortex-breakdown bubble region. This is the major stagnation area although there are other less dominant quiescent zones that have been detected by other vortex-breakdown researchers. If the stagnation zone flame goes out the whole flame would extinguish. Thus this is the key region of combustion from this point of view.

The rate of flammable gas supply to the stagnation zone may influence the size of the flame. This may explain the rapid changes in the diameter of the ball lightning. For an experimental vortex, in the non-combusting state, the strength of the down flow along the central axis possibly determines the extent and strength of the stagnation zone. In vortex-breakdown experiments in water it is found that the strength of the down flow is increased by increasing the swirl ratio. In a natural vortex this may easily change.

The combustion experiment demonstrated that it was possible to produce a flaming vortex using a readily available piped source of gas into the stagnation zone. This situation is clearly artificial and did not simulate how a natural vortex might sweep in gas into the burning zone. Also a natural vortex is unconfined while the vortex I produced was confined artificially and angular momentum was imparted to the air mass by a set of vertically mounted vanes.

How does a natural vortex suck in a fuel and burn this under natural conditions in the earth's atmosphere? I can only speculate here using ideas from the combustion experiments. Somehow a fuel gas must travel to the stagnation zone and be within the flammable range. So either the vortex accumulates the fuel gas inside the vortex from sub-flammable gas concentrations outside the vortex or the vortex funnel draws up high concentration gas. The latter case seems to be the better option. The gas volume concentration would only need to be a few percent because the lower flammable limit is in this range (eg propane 2-3%) The UFO photographed by a news reporter called Campbell near Sherman Texas in 1965 (see Part II of this book) shows the presence of small luminous zones actually in the funnel below the main breakdown zone. These zones are most likely low velocity eddies where gas is trapped and burnt. This suggests that gas coming into the vortex is indeed in the flammable range.

When I considered this option there seemed to be the conceptual problem that the burning vortex-breakdown moving through a combustible gas may actually ignite the whole gas cloud. Since this does not seem to be observed there must be an explanation. The reason could be that because of the high speeds the air and gas around the envelope of vortex-breakdown will not ignite. It is only when the gas enters the back of the vortex-breakdown burning zone that combustion takes place. The vertical down draft of air along the vortex axis also brings in combustible gas. The direct heating effect from the glowing flame cannot ignite the surrounding high speed air and fuel gas.

Consider the former possibility. Can the whole vortex embedded in a sub-flammable cloud of gas the vortex draw in and concentrate the gas into the flammable range and so combust the available fuel gas? If parcels of air and fuel are brought into the funnel there is, in theory, no net increase in fuel gas volume concentration. Some physical process would need to operate to separate the fuel gas from the air inside the stagnation volume and so lead to a build up there. I have not observed any such process in experiments.

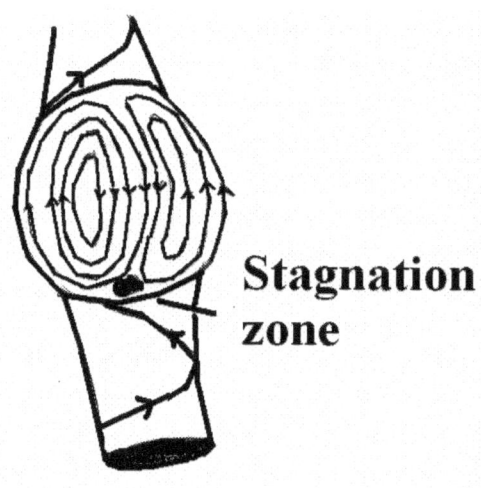

Stagnation zone

Figure 12: The stagnation zone is important to the vortex fireball theory since this is where the flame survives amidst the surrounding high air speeds of the vortex.

It is therefore likely that gas is not brought within the flammable range. The result is the vortex would not combust under these conditions. It might seem conceptually that gas could possibly be concentrated inside the vortex if this acted like an eddy it might trap the gas there leading to a fuel gas concentration build up. Experiments with water vortices reported by Leibovitch and Faler in 1977 have demonstrated that dye injected in the fluid flow can be retained twelve or so seconds within vortex-breakdown. The stagnation region where this dye was retained was an axial filament structure called the "inner cell". But this was not a demonstration of the increase in dye concentration in this zone since the dye was initially uniformly distributed throughout the vortex-breakdown. What had happened was that the inner cell of very slow moving fluid took a slower time to release the dye tracer.

Comments on modeling

More sophisticated mathematical models of these fireballs would involve the current results of vortex breakdown and flame research. A numerical combustion model which is a branch of computational fluid dynamics (CFD) could model the burning aspects. The dominant idea is that combustion inside vortex breakdown will unify and guide the explanation of a whole host of previously unexplained fireballs in the earth's lower atmosphere. Another mathematical model might use the turbulent Navier Stokes equations for unconfined swirling flow in vortex-breakdown a tornado-like vortex. This is a complex situation since there is a coupling between the equations governing turbulent flow and the actual combustion reactions that alter physical properties such as viscosity etc and the air flow pattern in the isothermal case. The paths of burnt and unburnt gases would also need to be mapped.

Support for the theory from scientists

After I had completed the experimental demonstration I wrote a paper and posted a copy to Dr Stanley Singer of Athenex Research Associates, Pasadena, California. He is the President of the International Committee on Ball Lightning (ICBL). He kindly arranged for Professor Karl Nickel to present my paper at the symposium in Kent. Karl is a German scientist who has worked in the field of gas dynamics. I was encouraged by his positive feedback from the symposium regarding my paper. Dr Singer, in an extract from a letter from the ICBL, dated 13 September 1995, reads:

"Both Prof. Nickel from Germany and Dr. M.B. Pankova, who is group Head at the Russian Research Institute of Aviation Systems, expressed their approval of your concept of ball lightning. The comment is especially gratifying since several quite different models were proposed by other Russian scientists at the Symposium."

When the 5th ball lightning symposium was advertised for Tsugawa, in August of 1997, I wanted to present my paper and meet with other ball lightning researchers such as Dr Singer and Professor Ohtsuki, the organizer of these international symposia who saw ball lightning when he was a ten years old. There were other personalities in the ball lightning area including; Associate Professor Ofuruton, the prolific Russian ball lightning researcher Professor Smirnov, the Dutch physicist and ICBL secretary, Dr Dijhuis, and the rotating dipole theorist, Dr Endean.

5
Japanese Fireballs

Y.H. Ohtsuki, H. Ofuruton [2] obtained information that BLs in Japan are often observed in good weather but in a period of seismic events.

[Quote from a joint paper presented to the fifth International Symposium on ball lightning held in Japan by Amirov *et al* from Institute for high Temperature-Russian Academy of Science, Moscow, Moscow Radio Technical Institute, St Petersburg State University.]

THE FIFTH INTERNATIONAL SYMPOSIUM ON BALL LIGHTNING

The International Committee on Ball Lightning (ICBL) organized a symposium in the Japanese town of Tsugawa renowned for "fox-fire", from 26 August to 31 August 1997. This international forum on ball lightning research was an opportunity to present my hypothesis to scientists. The organizers published the Tsugawa proceedings which was not done for the 4th International Symposium of ball lightning held in Kent. The countries represented included Belgium, Austria, Germany, USA, Britain, Holland, Japan, New Zealand and Russia.

I submitted an abstract that incorporated new applications of the hypothesis to other unexplained lights. These new ideas for the paper arose while I was writing this book and I used them for the Fifth Symposium. The vortex burner hypothesis was capable of explaining the Hessdalen lights phenomenon, black ball lightning, and other previously unexplained phenomena, which I will discuss later in Chapter 6. Much of the material for the paper originated from the Fourth Symposium. I wanted to have the paper published in the proceedings as a record of my exploratory work in this area. I acted for the Chemical and Process Engineering Department of Canterbury University, through Dr John Abrahamson. The abstract was accepted. I then set about writing and posting off a complete paper.

Tsugawa hosted the Fifth International Symposium on ball lightning. I was entranced and intrigued by the beauty of Japan and its people. Tsugawa is a tiny exotic mountain town in the heavily forested mountain country between the cities of Niagara and Koriyama. This locality is in striking contrast to the very first international ball lightning symposium which was held at the prestigious Waseda University in highly populated Tokyo.

Paper presentation

Thursday was the day of my afternoon paper presentation. There was also a public meeting on ball lightning for residents of Tsugawa, which included a Japanese film on the fox wedding parade, followed by a slide presentation by Dr Singer on ball lightning. In the evening there was a panel presentation of eyewitnesses who had seen mysterious lights or UFOs in and around Tsugawa. This was filmed by a national Japanese TV channel.

However my main job was to convey to the delegates that my ball lightning solution was the best theory currently available. I pointed out to the delegates that the theory had the ability to explain the current ball lightning observations, including a certain type of UFO. I maintained that the vortex fireball theory was a solution consistent with many of the qualitative observations of ball lightning in the literature. My line of argument was to suggest that if balls of fire have been observed embedded along the main axis of tornadoes then I saw no reason why scaled-down versions of these vortex burners could not exist in the form of ball lightning. I made it clear that the strategy of addressing the puzzle of the mysterious lights in tornadoes could yield a fruitful solution to the ball lightning problem. This was the reverse of the proposal that (Vonnegut, 1960) suggested. He said that in order to understand the unusual lights seen in tornadoes one might then have to solve the ball lightning problem first. The paper was well received with three questions from the floor which I answered from my knowledge of vortex-breakdown.

It may have been more constructive for researchers to discuss the strengths and weaknesses of the various theories presented at the symposium at special session since some of the ball lightning theories have little relevance to the problem concerning the nature of ball lightning and its published observations. For example a German experimental physicist named H. Schmidt-Böcking, who I met under a tree at a hall, one balmy evening, wrote a paper for the symposium in which "hollow" atoms were discussed (Schmidt-Böcking et al 1997). Yet as I read through his paper, co-authored with eight of his colleagues there was no link at all from the hollow atoms idea to the problem of ball lightning. More seriously I could not even find the word "ball lightning" in their paper and the hollow atom paper was a rather dubious sideways transfer from their own research area to the problem of ball lightning.

Schmidt-Böcking was not the only scientist at the symposium to apply their own field to the problem of ball lightning. There was the cold fusion specialist named Takaaki Matsumoto who gave a paper on ball lightning. Not surprisingly he linked ball lightning to cold fusion. Cold fusion science is obviously alive and kicking in Japan. He claimed to have produced tiny forms of ball lightning under water during electrical discharges using metal electrodes into electrolytic solutions of potassium carbonate and potassium hydroxide (Matsumoto, 1997).

A nonlinear specialist in dusty plasmas, Kikuchi (1991) wrote a paper entitled *"Electromagnetohydrodynamic vortices and corn circles"*. He conceived these so-called "EMHD vortices" as arising out of small scale turbulence. This work was a direct transfer from his own work in dusty plasmas. He speculated that these EMHD vortices could explain the English corn circle phenomenon.

There was little in the way of critiques of the various theories of ball lightning at the symposium, but there were two notable exceptions-Emeritus Professor Jennison and Dr Geoffrey Endean. Professor Jennison has some experience with ball lightning having seen it twice inside two different aircraft! He is skeptical, of some of the more esoteric ball lightning theories, including the cold fusion proposal. He said to me that the solution had to be simple-he did not go in for complicated and unlikely theories. I overheard that he had almost won a Nobel Prize. Professor Jennison, a radio astronomer, said some of his own ideas were conceptually very simple- it was just that no one had ever thought of them before. He said he had invented a simple but effective way of transmitting radio signals for amateur radio buffs.

Kitsune-bi or fox fire lights

On Thursday evening Professor Ohtsuki and Associate Professor Ofuruton organized a public meeting where several eyewitnesses were brought to the symposium building to recount their experiences of the mysterious lights called by the Japanese, "Kitsune-bi" or "fox fire" (kitsune, being the Japanese word for fox, and bi, fire. I first encountered stories about these lights from a senior policeman called Toshiki Kaneko outside the Koriyama Railway Station. He looked up a Japanese dictionary and found that they were blue lights seen in Japanese graveyards. But when I talked to Professor Ohtsuki over breakfast he said that the same mysterious lights seen around Tsugawa were all cases of St Elmo's fire. I was skeptical about this assessment, having discussed these lights with an American woman living in Tsugawa. She was working as a primary teacher and an interpreter. She said the kitsune-bi lights were basically spherical and they had the unusual ability to dart around in complicated motions. In my opinion this dynamic behavior did not put the lights into the category of St Elmo's fire. I have read elsewhere that the fox fire is the Japanese equivalent of the English, "will o' the wisp", an unusual flame that hovers in the air, and can perform daring acrobatics over the hedgerows.

My stance on kitsune-bi's nature was that the anecdotal evidence of unusual motion was inconsistent with a simple St Elmo's fire effect. St Elmo's is a corona discharge which stays in one place or if it does move at all it would remain attached to a fixed conductor. There is another reason because St Elmo's fire is a coronal discharge which is seen in stormy conditions while kitsune-bi can take place in fine weather.

In the early evening, after the symposium presentations, a public session was given which involved nine eyewitnesses of the mysterious lights. There was apparently great national interest in the phenomenon and these eyewitness accounts because the event was filmed by a nation-wide Japanese television station. The symposium scientists were each given English translations of the stories of each of the eyewitnesses. I was lucky to have the assistance of a Japanese interpreter. She confirmed that what was written by the eyewitnesses was a correct and faithful rendition of the actual talks. The eyewitness accounts were compiled and translated into English by Makoto Egawa (Egawa, 1977). As I listened to the reports I became more and more convinced that they pointed towards something that was more akin to my ball lightning hypothesis, rather than to St Elmo's fire. The mountain across the Tokonami River, named Mt Kirin, is a site that the locals frequently refer to as a "kitsune-bi" locality. I accept this local interpretation, but unfortunately in only one witness account given at the special public meeting, was a mysterious light sighted there. The mayor of the town was emphatic that these lights were a rather frequent occurrence about sixty or more years ago. The fox ceremony apparently originated from reports of these lights that were said to actually appear up on Mt Kirin. There were other reports in the general area of near Tsugawa and even further a field.

Bearing in mind that the witnesses saw the fireballs sometimes up to six decades ago, witness 2, called Tokoki Yamamoto from Tsugawa town, described how years ago in early autumn he saw, at around 9.00 pm, strange lights similar in form to lanterns mid-way up Kirin mountain. The location of the sighting corresponded to the location of a former castle which is now a tourist observation deck. The five to seven lights were strung out in a horizontal line and shifted in a direction upstream of the Tokonami River like a lantern parade. The lights were switching on and off and then suddenly vanished. Lights hovering in graveyards are a common occurrence in some cultures. In Japanese graveyards ghost-like lights have been seen hovering over the remains of ancestors. Susumu Yamauchi who lives at Yasuda near Tsugawa wrote:

" *It was 31 years ago in September when I witnessed it...it happened around 11.30 pm. I was looking outside the window, without paying much attention, and then a lantern like light appeared. First I thought it was a lantern, but then I looked more carefully, and noticed it was a sphere with irregular motion... The object was about 20 cm in diameter shaped like a sphere. I had thought that ball lightning moved in a straight line, but this moved irregularly, so I was scared, but at the same time interested. Before the object disappeared it flew straight up, made a circle, and then moving in the direction of the rice field disappeared. The next day, one of my students came to me and told me his grandmother had passed away. I found out later, that the direction in which the object disappeared, was in the direction of the cemetery*

[From an unpublished compilation of sightings presented at an International Symposium on ball lightning conference by Makoto Egawa]

The traditional association of these lights is connected with lantern-like lights which have been seen moving down a mountain called Kanabera Mountain, opposite the town of Tsugawa. This mythical tradition has it that these lights represent lantern lights carried by foxes as a wedding ceremony makes its way down the mountain. Even today the women in the town paint their faces white in imitation of this fox wedding procession. Yauko Dohi gave this testimony.

"I witnessed a kitsunebi (mysterious fire) in the fall of 1946. It was right before sunset and getting dark. Suddenly, there was a lot of noise coming from outside. I was curious and went outside to see. Then at the base of Kanabera mountain, lights the color of a lantern made a horizontal line, and repeatly appeared and disappeared. Then lights gradually became hidden by the mountain ridge and were gone. There were many other people witnessing this with me. This phenomenon, which lasted 20 minutes, reminds me of the story my mother told me about the kitsunebi. The fairy tale came true. It was extremely mysterious". [Makoto Egawa compilation]

The "lantern parade" description in the above sighting fits in well with an annual Tsugawa festival called the "Fox Wedding Ceremony". This ritual may have developed from the observations of strings of mysterious lights which could easily have been perceived as a lantern parade. Legend has it that whenever kitsune-bi is seen they have actually witnessed a fox wedding in the woods. In the town of Tsugawa the ceremony a young Japanese woman takes center stage. Her beautiful face was adorned with pale white paint to resemble a fox and an entourage of followers walked in a parade up the mountain.

A sighting of kitsune-bi was made by witness 8, named, Kii Takahasi, who saw a yellow-colored globe complete with a long tail, on a river bank of the Ubado River, which is a small tributary of the Tokonami River, flowing near the Isegu Shrine.

The mayor of the town, Mr O.Sawano, invited a kitsune-bi expert Yoshiharu Tsunoda to speak. He maintained that the people in Tsugawa do not see these lights any more. He said he did not know why there was a sudden cessation of the lights about 60 years ago, but he thinks it could have something to do with pollution in modern times. He believed the air and water have become unclean. But this explanation, as it stands, does not provide a direct and plausible mechanism to explain why the lights do not appear now. He also made the suggestion that the modern town lights have become sufficiently bright to outshine any unexplained lights that might emerge from the slopes of Mt Kirin. Unfortunately this light pollution suggestion is certainly open to dispute. I noted that symposium delegates easily saw mock kitsune-bi lights from Tsugawa at night time despite existing "light pollution" from the town.

After hearing the eyewitness testimonies of the kitsune-bi lights I was firmly of the opinion that if the lights could be explained using the vortex burner theory, it may be that the Mt Kirin area was a source of seepage of natural gas or hydrogen sulfide gas at the time of the sightings-some sixty odd years ago. Apparently for some geological reason the gas emission ceased although I did find circumstantial evidence for volcanic activity. I was told of the existence of hot thermal springs in the area very near to Mt Kirin. This geothermal feature is consistent with the possibility that there may have been a build up of gas on a calm night from gas seepage. Since Japan is a rich source of natural gas, it is conceivable that gas could have streamed up from some fissure on Mt Kirin, some sixty or more years ago. Natural gas survey maps of the area would help in this regard.

The possibility that such gas seepage in the area might still be taking place could be checked out by a geological field survey, perhaps with a gas detector. There may be other tell-tale signs that gas has seeped through cracks in the past. Such work may already have been done by Japanese geologists. In any case any evidence of gas emission would support a vortex burner theory explanation of these lights.

Just on dusk and before the banquet we were told by our hosts to look across the river to the dark silhouetted form of Mt Kirin. As we gazed across the dark form of the river we beheld a startling series of white lights dancing high up on Mt Kirin. Our Japanese hosts, In a skilful reconstruction, arranged for a group of men to wave bamboo sticks with a petrol-doused cloth on the top to act as flaming UFO lights. Thus the mystery lantern lights were recreated on the face of Mt Kirin.

Last day of the symposium

On Friday I had a conversation with the Russian scientist, Dr Marina Pankova who was to give her presentation in the morning. As we traveled in the minibus to the symposium hall, she showed me a book on Russian laboratory work on ball lightning, which was made available at the

4th symposium. Her work was on supersonic air flow with erosive discharges. She was positive about my theory, and said that she had cited my theory in a journal called, *"Chemical Physics"* which was one of the references she gave in her paper published in the proceedings (Leonov, Pankova, and Tomilov, 1997). She mentioned that Dr Singer considered my theory to be closer to the truth than many of the other ball lightning theories, especially the Russian theorists who presented papers at the 4th Symposium in Kent. I then told her that Dr Singer had written a letter to me citing how she and Dr Karl Nickel both approved of the vortex burner theory.

On the final day I asked Dr Singer (while we were eating breakfast at the Yasuda Western style hotel), how it was that he became involved in ball lightning research. Stan's reply was that he had accidentally overheard two very famous physicists, Feynmann and Pauli, discussing ball lightning in the library at Caltech. Feynmann said words to the effect that the existence of ball lightning was a figment of the imagination. Apparently the subject of discussion centered on ball lightning, and in particular, the microwave theory of Kapitza. Upon inadvertently hearing this conversation Stan was naturally curious to find out more. He then read an article about Kapitz's ball lightning theory and decided to have a go at analyzing the problem mathematically as a plasmoid in a gravity field. In his characteristically self-effacing manner, he considered the paper was a very modest attempt. He said he probably would not have got away with publishing his paper in that journal today. He abandoned the plasmoid approach and instead decided it would be more productive to review the ball lightning problem instead. Through US Navy funding he was able to write his now famous book which was completed over a number of years and published by *Plenum Press*.

6

Ball lightning explained

EXPLANATORY CAPABILITY

In this chapter I examine the ball lightning phenomenon as expressed in the literature. The vortex burner hypothesis unifies many observations under one explanatory framework while many competing ball lightning hypotheses lack this ability. The theory is also robust enough to be able to explain the diverse properties of fireballs like ball lightning, including large variations in such quantities as: lifetime, energy, size, shape, shape changes, motion, color and color changes and so on. I am making no attempt here to construct a mathematical description but to explain how the theory handles reported ball lightning properties qualitatively.

In a paper presented to the First International Symposium, at Waseda University in Japan, Barry and Singer (1988) proposed a simple test any proposed theory would need to pass to be a viable ball lightning theory. The test consisted of three basic observations that the theory should be able to explain. These are; (a) the long lived luminosity, (b) the occurrence of both regular and irregular motion, and (c) the spontaneous sudden appearance of ball lightning in clear weather.

The vortex burner theory satisfies the above three criteria because the long luminosity of ball lightning can be explained by a combination of a long-lived vortex and a continuing fuel supply. Secondly, the theory explains the regular and irregular motion of ball lightning through the typical translational motion observed of natural vortices, like the dust devil. Such vortices can move with or against the wind in seemingly smooth paths or more erratically. Thirdly, a burning vortex does not need a lightning storm. All that is required to create ball lightning in this theory is a high temperature source to ignite the combustible gas, a combustible fuel, and a vortex undergoing axisymmetric vortex-breakdown. In principle, fair weather sightings of ball lightning can be observed. Therefore, in principle, the hypothesis passes the Singer-Barry test. Other ball lightning theories, especially those totally dependent on lightning, would obviously not meet Barry and Singer's stringent test.

THE IMPORTANCE OF RELIABLE SIGHTINGS

The basis for both the existence and properties of ball lightning, and hence any promising theory of ball lightning, stands or falls on the authenticity of the observational evidence. It is realistic to expect that there will be some ball lightning sightings that are unreliable. The principle that credible and trained witnesses tend to give more objective reports is certainly valid. But even then, there are problems, as Charman (1981) has pointed out. Properties such as the true color, as opposed to the perceived color, can be incorrectly reported. Also energy estimates can also be erroneous. Smirnov (1987) remarked that a wildly inflated estimate of ball lightning energy of 428.4 kJ was made by a journalist, called Balyberdin, in the 5th July 1965 newspaper of "*Komsomol'skaya Pravda*". Such a figure managed to be included in the influential publications of both Singer (1977) and Barry (1980). According to Imjanitov, it was falsely assumed that the destruction of the house was directly attributed to ball lightning. This led to an overestimate of the energy density of ball lightning. Apparently the house's demise had nothing at all to do with ball lightning.

Some ball lightning researchers have applied statistical analyses hoping the averaging process would smooth out any anomalous discrepancies. Smirnov (1987) constructed a "mean" or "averaged" ball lightning with what he believed were typical field parameters. The main draw back of his approach is that it side steps the need to explain extreme values of certain specific parameters, like the diameter of the ball. Not all ball lightning events report the often-quoted average diameter of between 15 and 20 centimeters. Diameters much smaller or bigger than this have been reported. For instance, ball diameters in the order of several meters have been reported.

Jennison (1997) pointed out that statistical averages of ball lightning parameters could conceal large numbers of unreliable cases. He believed it was important to consider only the really reliable cases. But the problem is how does one determine what observations are completely reliable? One suggestion is to use the surroundings as a reference frame so that the diameter of ball lightning could be more reliably estimated.

CREATION OF CLASSIC BALL LIGHTNING

Much attention has been focused on what I termed "classic" ball lightning. Underlying this view is the idea that ball lightning must be small (20-30 cm diameter) and lasting a few seconds, at the very most, and always associated with storm and lightning. This is too narrow a conception as I have already pointed out. Nevertheless how could ball lightning be created under this narrow set of constraints?

Not every lightning stroke is necessary for creation. According to the theory described here, since only what is required are those lightning strokes that both create the vortex-breakdown and ignite gas in an existing vortex. For a lightning stroke to produce such a vortex it would to need to be a special one imparting vorticity, or rotational momentum to a stationary fuel gas-enriched air mass. This is a separate problem that could be addressed analytically by theory and experiment. The other way a vortex could produce a vortex is via a small localized gas explosion setting a mass of air spinning in air mixed with a fuel gas. This suggests an experiment using an artificially-triggered lightning stroke (or from an artificial high voltage source) discharging into a natural gas inside a container with various instruments such as pressure sensing instruments. This could end up as a transient (unsteady) spinning, burning, air mass (ball lightning) but would be harder to study than under more controlled conditions such as the combustion fireball I produced in the steel vortex generator within the Chemical and Process Engineering Department. I believe that the latter method is a more productive and cheaper way to grasp a better detailed understanding of the swirling flame in a tornado-like vortex.

LIFETIME OF BALL LIGHTNING

One common misconception about ball lightning is that it is a short-lived phenomenon with a typical lifetime up to about one second. Such a time is not in agreement with published lifetimes. For example, in my master's thesis, I cited Grigorjev et al (1988) who reported 800 instances where ball lightning lasted longer than 12 seconds. Many other surveys reveal lifetimes greater than 1 second. Smirnov cited four reviews with the following median lifetimes: McNally (1966)-4 seconds, Rayle (1966) and Charman (1979) both 5 seconds, and Stakhanov (1979)-14 seconds. Barry (1966) reviewed over 400 cases and 83 percent of his reported events had lifetimes of 5 seconds or less. A smaller percentage of these were longer-lived ball lightning events were attributed to another cause-St Elmo's fire.

The vortex burner theory is ideally suited to explaining long lifetimes because natural vortices themselves have wide ranging lifetimes. Furthermore it could be that the combustion process may assist in extending the lifetime of the vortex by providing an additional source of buoyant up draught. Certainly it has been demonstrated that an electrical discharge can sustain a vortex. So although this idea needs further experimental and theoretical work, the principle looks sound. The lifetime of ball lightning, if the vortex burner description is valid, would logically depend on two critical elements. The first is that the vortex must sustain vortex-breakdown, and secondly, there must be fuel gas or combustible, in sufficient concentration and quantity for sustained combustion. If any one these conditions are not satisfied the vortex flame would not survive.

There is documented evidence of ball lightning appearing to repeatedly disappear and then reappear. Such an observation is easy to explain and there is no reason why a vortex could not, in principle, become luminous a number of times, provided the gas entrained in the vortex could be either rekindled with new fuel or existing fuel re-lit by an ignition source.

Does burning extend the lifetime of the vortex fireball?

Let us explore further how the burning flame within vortex-breakdown might interact with the overall air flow to prolong the lifetime of the vortex. Swirl burner studies have shown interactions between the flame and vortex-breakdown where vortex-breakdown tends to stabilize a flame. The same could be true in natural vortices. Vu and Gouldin (1982) found in their industrial prototype swirl burner that in the special case of pre-mixed reactants, the recirculation region of vortex-breakdown helps to stabilize the combustion flame. But this work does not say anything at all about whether burning would provide a sufficiently strong up draught to prolong the lifetime of the vortex.

Rapid combustion can generate an updraft and a vortex is already observed in nature in the form of fire whirls. Fire whirls are vortices produced as a direct result of the rapid combustion of fuel such as petrol or in forest fires. But fire whirls are not vortex fireballs; they are a low swirl phenomenon. The fire whirl phenomenon is quite different conceptually to the vortex burner theory, because, in the former case, most of the combustion is not specifically contained within in vortex-breakdown. The fire whirl is formed from a convective updraft in the immediate environment. The fire whirl typically develops into a long vortex core embedded in these flames.

Ryan and Vonnegut (1970) showed experimentally that a high voltage electrical discharge can create and maintain vortices in a rotating mesh cage chamber. In this case electrical energy is converted into heat and continues to generate convection currents. There seems no reason why a combustion flame could not substitute for the electrical arc discharge to sustain vortices from buoyant up draughts.

I found by experiment that provided there is a vertical air flow, (mechanically driven air flow using a fan), and a steady source of vorticity (the metal vanes) the vortex-breakdown will persist. Obviously for weak vortices the source of vertical convective air flow will not last, even though the source of vorticity may still be present. So it seems logical that the combustion process of a burning vortex could assist in extending the lifetime of a freely moving vortex in the atmosphere. With a steady input of fuel gas to the combustion zone perhaps the vortex fireball could remain for longer periods of time, especially for larger vortices that last several hours.

ENERGY AND ENERGY DENSITY

Far from being a passive floating luminous sphere, ball lightning is on record as being associated with large amounts of energy. Ball lightning has been known to dig trenches, lift objects, and knock over objects (Singer, 1971). This is exactly what large vortices can do. They have been known to "dig" or "drill" holes and even trenches in the ground as well as lifting heavy objects into the air.

I found references to fair-weather, short-lived "mini tornadoes", whirlwinds and "twisters" that have displayed the same sorts of mechanical effects as ball lightning and with a similar lifetime, during fair weather conditions. The *"Christchurch Press"*, of 7th January 1997 reported on whirlwinds at Lake Brunner. Karen Askew of Iveagh Bay was sitting on a beach, around 3pm Sunday, when a whirlwind came down and knocked some children over. The whirlwind lasted about a minute and created a pattern in the sand. A second twister was seen at 12.30 pm and lasted for 3 seconds.

Energy density

There are several ways to quantitatively assess ball lightning energy. One could look at the power output which is the total energy of the ball over its lifetime. Then there is frequently quoted parameter of energy density. The energy density is defined as the total energy contained in the ball at the time of creation, divided by its volume. Researchers have put the lower and upper limits of energy density from 0.01 Jcm^{-3} to 1000 Jcm^{-3} (eg Smirnov, 1987).

Natural gas is the most likely candidate for the fuel gas powering ball lightning and has a heating value of around 37.5 MJm^{-3} (Jones, 1993). Even higher values are possible for natural gas with greater proportions of ethane. Jones quoted a case of a high ethane containing natural gas fuel with a value of 58 MJm^{-3}. These values have the same units as energy densities but they actually represent the heat output of the gas if one cubic meter is burnt under standard pressure and temperature conditions. This type of chemical "energy density" is different to the type of energy densities that some ball lightning investigators calculate which are much higher.

I believe the reason for such abnormally high energy densities is the assumption that all the energy is packed into a single 20 centimeter ball at the moment of creation. These calculated high energy densities, I believe, can be accounted for by making the assumption that the total energy is supplied externally to the ball mainly through chemical energy over a given time interval divided by the volume of the ball.

Energy density of the Goodlet fireball

To account for the anomalously high energy density in one well known ball lightning event I will use the above two values of the heat output of natural gas per cubic meter to obtain a numerical value of "energy density" with the same order of magnitude as the classic case of ball lightning reported by Goodlet (1937). A fireball with a 15 centimeter diameter dropped into a barrel of water containing 4 gallons of water. The water boiled for 4 minutes. Energy density estimates taking into account water evaporation were calculated. Barry (1967) calculated the energy density as 2500 Jcm^{-3} while Smirnov (1987) put the estimate as between 2000-6000 Jcm^{-3}.

Consider a 15 centimeter diameter ball lightning. Assume for simplicity that the combustion takes place in a spherical volume with this diameter. Assume there is a complete transfer of heat energy from combustion to heat the water and evaporate it in the barrel. Take the vortex core leading into this fireball as one third of the diameter-this is about right for vortices undergoing vortex-breakdown. As a rough estimate of the flow rate into the fireball consider only the axial component of air/fuel gas flow up the vortex core. If the axial speed of the air-to-fuel mix is around 5-15 ms^{-1}, which is a reasonable estimate for a small vortex, and using natural gas fuel with a calorific output of 37.5-58 MJm^{-3}, a diameter of eight centimeters (orange-sized) and the reported time of 4 minutes, it is possible to get, using a spreadsheet sheet analysis, energy density estimates of around 2 kJcm^{-3}. This figure is the same order of magnitude as the energy density estimates of Barry and Smirnov for the Goodlet fireball- i.e. between 2000 to 6000 Jcm^{-3} respectively. Thus the high energy density estimates can be explained on the basis of the heat output of natural gas over the observed time interval.

Does adding the vortex mechanical energy further increase this energy density of the Goodlet fireball? A rough estimate of the power of a moving air in the vortex was calculated by Vonnegut (1960), as $P = 3.14pr^2V^3/2$, where P, p, r, and V are the power, density of the air (~ 1 kgm^{-3}), radius of the vortex core, and the average speed of the air mass respectively. Energy is equal to power multiplied by the time. When I carried out the calculation I found this "additional energy density" to be negligible compared with the chemical energy density. I used 5ms^{-1} for the air speed but even going up to 10 ms^{-1} did not significantly alter the result.

MECHANICAL ENERGY

A vortex, like a tornado, is well known to exhibit strong mechanical effects such as the lifting of heavy objects and the digging of circular holes and long trenches into the ground. Are there cases where ball lightning does the same thing? There is evidence. There are specific ball lightning events reported by Corliss. On 6th August 1868 a two foot diameter red fireball was seen for twenty minutes and reported by Mr M. Fitzgerald in the Glendown Mountains in Ireland. It dropped from a ridge into a valley and hit the peat bank of a stream to leave a huge hole and a very long trench (several meters). [I have another description of red fireballs in a lightning storm in County Armagh where the fireballs dug and burnt craters the size of basketballs. They were seen dancing on the road and are clearly not meteors!] The fireball reduced in diameter to three inches when it disappeared. In Essex, England around 3 a.m. there was a flash and an explosion and a ball of fire appeared.

"One person stated that he saw what appeared to be a cylinder, and another person, a ball of fire, descend and then explode, 'casting darts' in all directions. On careful examination in daylight it was found that an oatfield which had recently been dredged there were three distinct sets of holes ranging from 9 in. down to 1 in. in diameter. The holes were perfectly circular, diminished in size as they went downwards, and remained so onto the perfectly rounded ends at the bottom. Upon digging sectionally into the soil, which is stiff clay, it was found that the holes were 'as clean cut as though bored with an auger'."

There is another remarkable case of a fireball three inches in diameter cutting a trench ten feet long. This took place on the 12th July 1868 in Guilford, England and was reported by Capron in *Symon's Monthly Meteorological Magazine,* 18: 41, 1883. Another ball lightning case was reported in *Nature*. The fireball was seen on the 1st August, 1907 at Alpena, Michigan. It dropped to the floor and traveled around the room in circles smashing holes in the walls. On the 3rd of July, 1892 a 60 foot column of water was raised after a fireball dropped into a lake near Liverpool, England. The best explanation was that water raised was by a corkscrew motion of air moving air and water upwards into the vortex.

HEATING EFFECTS

Significant heating effects have been attributed to ball lightning sightings (Singer, 1971). This property of ball lightning seems to demonstrate that ball lightning models which propose low energy and virtually no heating effects are inadequate. It is on record that sufficient heat transfer has taken place in some ball lightning events to cause the burning or melting of objects. The boiling of water in a barrel is a classic case. (Goodlet, 1937). Observers encountering ball lightning have felt heat and in some cases they have been burnt by the ball. A 10 centimeter bright blue and purple ball (pre-mixed flame?) with a yellow halo (diffusional flame?) burnt a 4 by 11 cm^2 hole in the dress of a woman who observed it. Her hands and legs became red with the contact of the object (Singer, 1977)

According to the vortex-breakdown burner theory there should be a transfer of heat by conduction, convection and radiation to the immediate surroundings by the flame confined in the aerodynamic recirculation of vortex-breakdown. For a steady-state flame the chemical energy input from the fuel gas and air will be balanced by energy loss from heat transfer processes of convection, conduction and radiation. This steady-state flame could be achieved in the laboratory, but in nature the fuel supply can be variable resulting in an unsteady flame with corresponding changes in flame color and luminosity.

Much of the radiation emitted from a ball lightning vortex fireball is expected to be in the infra-red, if ordinary flame studies are anything to go by (Gaydon and Wolfhard, 1953). As far as an observer looking at the fireball is concerned, the bigger the fireball the more infra-red radiation flux will reach the observer, provided the distance of the observer to the fireball is kept constant, and other parameters such as emissivity and surface temperature are also fixed. Note that the emissivity is a numerical ratio of how an emitting body departs from black body conditions. A black body is a perfect emitter and absorber of radiation and so has an emissivity defined as 1. Flames normally will have an emissivity less than one. The so-called "shape factor" which is a term used in radiation theory, needs to be calculated if radiative heat transfers are to be estimated. The shape factor for radiation from a fireball sphere to a human body would include the fireball radius in well documented shape factor formulae.

One can appreciate the shape factor effect by imagining a bonfire of wood on a beach and noting how far away you would have to stand to be comfortable with the heat. With a small fire burning the same material you can need to move much closer to the fire to get the same heating sensation on your skin. Thus bigger spheres of ball lightning should be associated with greater heat flux reaching the observer. However the measure of this flux per unit time- the radiant power (kWm^{-2}), is dependent on the surface temperature of the ball, which for a grey body, (an approximation for a flame) depends on the fourth power of temperature through the Stefan-Boltzmann law. This would explain why heat is sometimes felt with ball lightning while in other cases heat is not felt. At lower flame temperatures (perhaps sometimes from "cool" flames of hydrocarbon fuels) the body being a poor detector of heat radiation may not be able to detect heat from the ball lightning.

By far the biggest heat loss for a vortex fireball moving in the atmosphere should be through convection, where cooling air is moving up, and around, and even into the fireball at the rear. The only reason why there is not rapid cooling is because of the continued gaseous fuel input to the fireball.

The more intense blue or blue violet balls globes should be connected with higher temperatures produced at or near stoichiometric proportions. Under these conditions a pronounced heating transfer will take place. Stoichiometric or near stoichiometric flames would be capable of melting of low-medium melting point metals and glass.

MAXIMUM TEMPERATURE

From thermodynamics theory the maximum flame temperature is reached when the forward and backward chemical reactions proceed at equal rates. This means that products are formed at the same rate as they are dissociated into reactants. Natural gas has a maximum adiabatic (no heat loss) flame temperature of around 2000 degrees Celsius. This is a theoretical upper limit to the flame temperature which is expected to be approached under stoichiometric conditions in well-insulated furnace conditions. Methane, usually the main component of natural gas, in many situations, has a stoichiometric value of 9.5% (Gaydon and Wolfhard,1953). However, even with an uninsulated or open flame system temperatures can approach the maximum theoretically possible. It is known that a methane gas welding torch flame, under stoichiometric conditions, can cut through steel, and even copper, if it is well insulated (Deev, 1998). It then seems feasible that a vortex fireball burning natural gas could melt its way through glass which has a lower melting point temperature than steel.

COLOR AND COLOR CHANGES.

All colors of the visible spectrum have been reported by observers of ball lightning. Thus we have red, orange, green, blue, indigo, violet, and a dazzling white color. Even cases of "black" ball lightning have been reported. There are even cases where ball lightning is multicolored and two-toned balls are not uncommon eg. blue-white or yellow-orange. Color changes have also been observed.

There is some difference in opinion as to which color is most frequently reported in the literature. Singer (1977) noted that red and orange were most often reported, while other reports seem to suggest a blue or blue-violet color.

One problem for ball lightning investigators is how to properly interpret reported colors of ball lightning. For instance, when a ball is described as "yellow" is it the yellow that would be normally associated with a yellow diffusion flame or is it an orange yellow connected with sodium emission from a flame? To some observers a "red" might mean a pink color which would place it at a higher temperature than the usual red color under black body conditions. This subjective element in the color assessment of ball lightning by an untrained observer could lead to incorrect conclusions about temperature of ball lightning and what the nature of the processes producing the color emission in the visible region.

Explanation of color.

The possible processes likely to give rise to ball lightning color can be qualitatively explained using the vortex fireball theory. Fundamentally, the color of luminous ball lightning represents visible emission of light from the flame. Non-luminous ball lightning events are vortices entraining visible material like smoke.

One important prediction of the vortex burner theory is that because the maximum flame temperature is around 2000 degrees Celsius then colors due to black body radiation in the green, blue and violet regions are ruled out. This is because much higher temperatures are required for these black bodies. Red and orange ball lightning flames might be possible colors corresponding to black body flame surface temperatures of around 800-1000 degrees Celsius. There is the additional problem that most small flames do not act as black bodies, although larger flames can become self-absorbing and approach black body conditions (Gaydon and Wolfhard, 1953). Even if the flame did approach black body conditions for larger cases of ball lightning, chemical equilibrium would be reached at much lower temperatures than the higher black body temperatures of 10 000 plus degrees Celsius for black body emission at the peak wavelength in the blue region.

An alternative explanation of flame colors needs to found for lower flame temperatures other than black body. When the flame is not stoichiometric then the flame should generally appear a yellow color, in the absence of other impurities, because of luminous emission from carbon particles present in the flame. Carbon particles in an optically-thick flame act as black body emitters with a maximum wavelength in the yellow part of the visible spectrum, which corresponds to a temperature of around 1400 degrees Celsius.

Normally for flame viability, the concentration of natural gas should lie between the narrow flammable limits of 5 to 15% by volume of gas. Though reports of combustion flames have been reported for fuel gas concentrations lower than the usual flammable limits. Any deviation from the stoichiometric concentration (for methane it is around 9.5%). will end up as a yellow ball lightning in the absence of any other impurities and with a natural gas fuel with a characteristically high methane concentration. But a certain fraction of vortex fireballs in the earth's lower atmosphere will be at, or near, stoichiometric concentrations and burn with a blue or blue violet flame. These blue ball lightning events will, according to the theory correspond to a hotter temperature than a yellow flame which has a lean or rich natural gas-to-air mixture. Since it is easy to conceive of varying concentrations of natural gas brought into an atmospheric vortex there will be a range of ball lightning flame gradations from total yellow to yellow-blue, to blue and blue-violet. In a turbulent pre-mixed flame there may be localized regions which vary from stoichiometric and so the ball could have several different regions of yellow and blue and violet.

Red or red-orange balls.

Some ball lightning events are reported where the fireball is completely red or red-orange. It is difficult to see how a uniformly red fireball can come about in the lower atmosphere if natural gas is the fuel gas. While red zones of pre-mixed natural gas and air flames have been observed by Gore (1998), it seems that wholly red natural gas flames are much rarer.

The color of ball lightning would therefore be either yellow or blue and sometimes violet with varying gradations, but not fully red. There may be other fuel gases which might be present in the atmosphere but I have not come across any common gas that might give a red color. And it would need to be a reasonably common gas to account for the high frequency of ball lightning of this color.

A number of possible explanations are possible and may resolve this problem. The red color could stem from the fine carbon particulate material trapped within the vortex-breakdown combustion recirculation region. This particulate material would be much larger in diameter than carbon particles that emit yellow light in candle flame. These particles would be large enough to be visible to the naked eye and glow red-hot like the embers in a fire or the charcoal on a barbecue. This might mean that the initial material might have burnt and produced the usual yellow flame color and then settled down to glow red. The red radiation would be essentially black body radiation from the glowing carbonaceous material, though the hot air around these particles which would be at a lower temperature. This would put the temperature around 600-700 degrees Celsius, depending on the type of redness. I expect the source of this carbon material would be mostly from wood or sooty material in the atmosphere or on the ground where the lightning strikes. Red flames have been seen in log burners which burn pine when the air to the burner is reduced substantially (Deev, 1998). Dark smoke was seen above the red flame indicating that these carbon particles were emitting in the visible red. The red flame was all red and appears to require the special conditions of the log burner. Such a flame in open surroundings could not survive. One reason only a small amount of air is required to keep the flame going is perhaps because the air is preheated. But it is suspected that with this reduced supply of air lower temperatures than the usual flame in a wood fire will result.

If this effect is to take place in connection with ball lightning, the vortex-breakdown must have some kind of medium around the flame front to reduce the air to the spherical flame, in an analogous way to the log burner. One suspect is soot. Beer and Chigier (1972) cited evidence that some fully rotating flames were highly luminous with large concentrations of soot but more importantly the oxygen concentrations that were measured were lower in the flame. Furthermore soot formation can be three to four times larger than steady laminar flames (Zhang and Megaridis, 1998). Since we are concerned here with the combustion of natural gas, the fluctuating flow of vortex-breakdown could manufacture soot manufacture within the flame in quantities higher than for ordinary laminar and steady flames. High residence times of particles in vortex-breakdown could also enhance soot formation.

There is also a quantum electronic transition process that could impart a red color to the flame. Metallic ions, usually from salts, can add color to a flame. For instance, if ball lightning is seen at sea the ball lightning flame, in the presence of sodium ions, will possess a characteristic orange-yellow color. If there is copper ions present in, say, atmospheric dust then ball lightning will appear green. Calcium salts have a characteristic red-orange signature color and potassium salts produce a violet color in a blue Bunsen flame. The color change of the flame is therefore not indicative of the chemical combustion reactions, but of the color due to certain atomic transitions from impurities present in the flame. This effect is well known in chemistry flame tests where metallic salts, such as copper sulfate, are introduced into a Bunsen flame.

If red globes are to be explained by this quantum mechanism then one candidate would be calcium salts which do produce an orange-red color. Particulates composed mainly of calcium compounds with low fall-out speeds might provide a source of impurities to color a vortex-breakdown flame red or red-orange.

Green fireballs

Green fireball flames could occur from metallic ions, like copper and nickel, which are transition metals that emit visible light at wavelengths in the green part of the spectrum. Reports of green ball lightning fireballs should be less common, than say blue, red or yellow fireballs. This is because natural gas fireballs would have to encounter airborne copper ions which are not as common as other impurities in the air. This prediction appears to be supported by Smirnov (1987) who combined four surveys and calculated an average value of 1% for green balls out of a total of 1467 sightings.

Another prediction is that green ball lightning would be observed in areas rich in copper ions. Green ball lightning will be produced from copper electrodes in the laboratory experiments or in areas rich in elemental copper or copper salts. The classic case reported by Silberg (1962) of green fireballs which floated off the reverse current electrodes of a relay switch in a submarine. The relay switch allowed charging to take place between a D.C. generator and two sets of batteries. Large currents were involved (156 kA at 260 V). The electrodes were copper and silver so the idea that the green color is to do with copper ion transitions cannot be ruled out.

White fireballs

Fireballs that appear white suggest a mixture of colors emitted from all parts of the visible spectrum. In reality some fireballs may appear white when in fact they are not. This is like the light from a tungsten filament which can seem to appear white but on close inspection is more like yellow. If the natural gas flame, under stoichiometric conditions, is mainly emitting blue light then it would require impurities in the flame to produce this continuous spectrum. The impurities may come from certain metals or compounds which glow like a hot gas mantle,

Some comments on blue fireballs

If natural gas is implicated in ball lightning sightings, then a sizable fraction of sightings should be blue. Several surveys bear this out, with the notable exception of Barry (1967), who used a survey containing about 400 cases. He said that even though there was a high frequency of blue observations in the early surveys of both Mathias (1934), and Brand (1923), the blue or blue-white color is a color incorrectly attributed to ball lightning.

In my opinion the earlier surveys would appear to confirm the prediction by the vortex burner theory that there will be a significant number of blue color sightings, perhaps suggesting that natural gas is the predominant fuel gas. Actually, even later surveys contain high proportions of blue ball lightning events. Ofuruton et al (1997), in data obtained by the Japanese Information Center of Ball lightning, reported 63 blue to violet balls out of a total of 231 cases (27.3 %). Eyewitnesses also report a blue color. Jennison (1997), a physics professor saw a deep blue ball inside an aircraft. This blue ball lightning did not emit any thermal radiation that was detected by the skin even when it was fairly close. This could be a case of a vortex fireball with a cool flame with the fuel gas in a low concentration.

An anecdotal report in (Cade and Davis, 1967) suggests blue ball lightning as being quite common in Norway. They mentioned a friend of theirs, called Kari, from Oslo, who said that Norwegian ball lightning called "rullende lyn", or rolling lightning balls was reported as being either blue-white or blue.

Why did Barry generally exclude blue ball lightning sightings? Barry believed they were not ball lightning, but either the ignis fatus (will o' the wisp), or St Elmo's fire. The will o' the wisp is itself unexplained and incidentally can move like reported cases of ball lightning. In fact the will o'the wisp may have the same nature as ball lightning (which is actually what I propose in this book.)

Although I share Barry's concern that a certain percentage of events previously classified as ball lightning may in fact be St Elmo's fire, there is a danger that by omitting all blue fireballs genuine cases of ball lightning might be ruled out and mistakenly identified as say, St Elmo's fire. Barry's criterion for an object to be St Elmo's fire was that the glow would need to be seen in thunderstorms above an earthed object. The lifetime is in the range of many seconds or even minutes. The color should be either orange or blue to blue white sphere or of an oval form. The diameter is bigger than 0.3 meters and that the phenomenon decays either suddenly or slowly but always quietly. St Elmo's fire cannot move independently, like ball lightning, which can fly in all sorts of trajectories, but is constrained to move in the vicinity of a conductor and cannot break free to roam around.

Color changes of ball lightning.

It follows from the theory that the mechanism for ball lightning to change its color stems from variations in the fuel-to-air ratio from the stoichiometric condition or by the introduction of different impurities into the flame. Thus a blue stoichiometric flame when suddenly entering a region containing copper ions would transform to a green fireball.

PREDICTING THE TYPE OF SPECTRA OF BALL LIGHTNING FIREBALLS

It follows from the theory that ball lightning should possess emission spectra which is consistent with previous studies of natural gas combustion. Thus for a blue white flame at the hottest temperature a spectral analysis should show the existence of emission species like CH, OH, NOx, etc as reported in flame research (Gaydon and Wolfhard,1953).

ELECTRICAL AND MAGNETIC EFFECTS

Studies of flames have shown that inside a flame there are electric charge carriers produced by the chemical reactions taking place. The main ionization in a flame is considered to be due to electrons from thermally-generated dissociation processes. The recombination (neutralization) of this charge is said to slow down at higher flame temperatures (Gaydon and Wolfhard, 1953). The charge density of hydrocarbons like methane can reach 10^{12} electrons per cubic centimeter. It has been established in experimental studies that there are additional negatively-charged carriers other than electrons. If this electronic charge is on the move, and does not rapidly recombine, then it is possible that a net electric current may be produced to generate a net magnetic field. In larger flames there will be larger amounts of electric charge and therefore large magnetic fields surrounding the flame.

Unfortunately clear evidence of ball lightning magnetic effects in the literature are very scarce, although ball lightning has been reported as affecting ships compasses (Corliss, 1982). However some reported magnetic effects may be connected directly with the lightning stroke than being created by the ball lightning.

The "Hall effect" has been observed in hydrocarbon flames which indirectly demonstrated the presence of magnetic fields associated with the flame. The effect occurs when an external magnetic field is applied at right angles to a flame. Moving electrons perpendicular to the applied field are pushed by the magnetic force to one side of the flame and positive charge to the other side setting up a small voltage drop. The mobility of ions can then be determined using this method. The Hall Effect may eventually be demonstrated in connection with ball lightning. This is another prediction of the vortex fireball theory. That the applied magnetic force can push this moving charge along could only happen because the moving charge itself has a magnetic field. That is, a moving electric charge produces a magnetic field.

Such ionization within the flame would be able to reflect or absorb electromagnetic radiation in the microwave or radio frequency range. Thus ball lightning should show up on radar if the flame ionization was present.

There are other electrical phenomena associated with flame ionization. External electric fields have been known to distort the flame because of the ionization present (Gaydon and Wolfhard, 1953). Such a flame distortion should eventually be detected in association with ball lightning. Any oscillation of this electric charge within the flame will also generate electromagnetic radiation.

Various tools such as radar and a Gauss meter could be used to measure the magnetic field around a given ball lightning to find out if it fits the predicted strength of magnetic field from such an ionization source. The degree of ionization could be measured using standard laboratory techniques like microwave attenuation. If microwaves were to be shone at the suspected fireball and the attenuation measured, then the electron charge density could be calculated.

SIZE AND SHAPE.

The diameter of ball lightning can range from 20 millimeters up to an amazing 20 meters (Singer, 1977). Why is the mean diameter of ball lightning often reported to be in the 20-30 centimeter range (Smirnov (1987)? This is the classic size for ball lightning. The vortex fireball theory suggests three possible scenarios. The first is that an average lightning strike generates a vortex by heating a column of air, and in doing so ignites the natural gas feeding into the vortex from some source. The second situation is where lightning strikes an existing gas-laden vortex, while the third case lightning detonates a natural gas/air mixture creating a burning vortex. In this case the fireball would then emerge into existence with an explosive bang.

A prediction from the hypothesis is the possible correlation between energy of lightning stroke and diameter of ball lightning. Bigger strokes could produce bigger vortices up to some maximum limit. The upper limit would be determined by the maximum energy released by a single lightning stroke. However, higher energy strokes tend to give larger vortices this would be offset by greater axial velocities which may produce stronger vortices but the lateral extent of the vortex would be similar to lower energy strokes. Another prediction is the existence of upper size on ball lightning set by the largest diameters of the biggest tornado vortices. Thus the diameter of ball lightning should range up to, around 30 meters. The lower limit of ball lightning's diameter would correspond to a few millimeters for very small vortices.

Ball lightning with tails

A tail structure above or below the fireball is a reasonably frequent observation. The Japanese data shows a strong preference for tails. Of 250 sightings selected from 2000 events, 39.80% were spheres with tails (Ofuruton et al, 1997). The luminous tail or spiral structure below or above the fireball section of a ball lightning is simply a luminous flame extending below the fireball of vortex-breakdown into the funnel or vortex core. Dust or water droplets could be sucked into the fireball and act as tracers to make the "tail" visible.

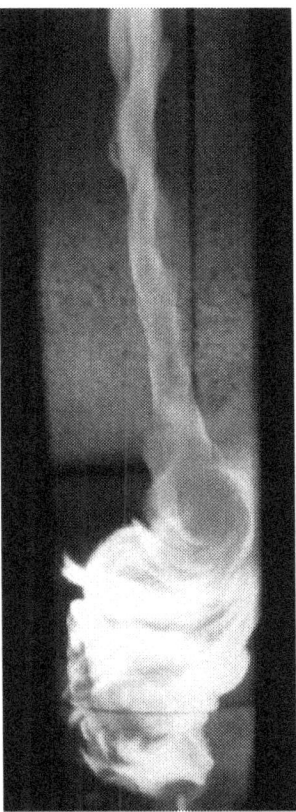

Figure 13: A high mixture ratio flame in the 0.9 m diameter vortex chamber. This work was reported in Coleman & Abrahamson (1997). This flame has the rope-like vortex above vortex breakdown and is observed in the well publicized Berger ball lightning and Brett Porter's ball lightning photograph mentioned earlier. Copy right prevents full reproduction of these images but they are widely available and I have indicated the sources.

Burger's vortex

A photograph of ball lightning with a cord-like tail and widely regarded as authentic has entered the literature through Encyclopedia Britannica (also the CD version). The published photograph is thought to be a genuine case of ball lightning photographed by in 1978 by Werner Burger at night at Sankt Gallenkirch, Austria. I examined the photograph and noted a rope-like structure, providing circumstantial evidence for spiraling air up the funnel. This is the same phenomenon I produced in the vortex chamber experiments (see photograph above). The cord structure is identified as the vortex core while the lower luminous region is flame combustion inside the vortex-breakdown region itself. It could be one of the best photographs of ball lightning available showing combustion in a vortex under natural conditions.

Explanation of shape and shape changes.

Ball lightning is not always spherical, as we have seen earlier. It can take on various shapes like a cylinder, ellipsoid, tear-shaped, donut and so on. Ball lightning structures are not always stable suggesting an ongoing process within ball lightning that can alter the shape. For instance, ball lightning can suddenly change from a soccer ball shape to an oval rugby ball form in fractions of a second. I propose that these changes originate mainly from the changing air flow distribution within the vortex. The vortex-breakdown recirculation air flow structure can alter because of perturbations of the air flow entering the vortex. These alterations in air flow transform the flame envelope.

It is worth mentioning that other ball lightning theories contain the possibility of explaining the shape changes to ball lightning. In a report on the 4th International Conference on Ball lightning in the January BL newsletter Dijhuis (1995) N.Kondo reported on complicated resonance modes generated in microwave cavities. It is unclear from this brief report whether these resonance modes correctly match up with the actual shapes of ball lightning observed.

The vortex-breakdown-combustion hypothesis can be used to explain some of the various shapes reported. This is because they can be derived from a consideration of the flame shape arising from the basic form of the vortex-breakdown bubble. The structure of ball lightning is variable, ranging from: basic sphere, spheres with trailing tails, spheres with a cylindrical appendage above or below or both, two or more balls linked vertically or separate, balls all in chain-like formation, pear shapes, cylinders, cylinders with conical ends, torpedo shapes, a ball shape with two tube-like appendages at the top, a torus shape, amorphous blobs, discs, balls with a spiral feature above or below, club shaped, hollow ball shape, vertical dumb-bell shape, ray-like projections pointing radially from a ball (usually seen just before the ball explodes). The torus shape may be related directly to the tori reported in the vortex breakdown work of Sotiropoulos & Yiannis (2001) and elsewhere. Residence times of dye in these regions were reportedly very long. A fuel gas could also remain in this zone and give it a luminosity.

The diverse ball lightning shapes reported are amenable to an explanation utilizing the basic "ball and tube" geometry. For instance a set of "cats ears" protuberance on top of a ball is actually a reported shape in the literature and could correspond to vortex-breakdown with the same kind of structure (see photo in this book). A basic guiding principle to interpret these shapes is to think of how the circulating air fuel patterns within vortex-breakdown bubble region interact with the combusting flame region, and how in turn, the flame alters the air flow. If industrial swirl burners are anything to go by, the flame length and shape will be influenced by the so-called "mixture ratio". I have found in my burning vortex experiment that the length and flame shape did change with the amount of fuel gas but the swirl angle was also important.

Why spherical fireballs?

Published and unpublished photographs of vortex-breakdown in the literature (eg. Coleman, 1990) demonstrate that a perfect spherical vortex-breakdown has not been achieved. So how is it that a near perfect spherical ball lightning is to be explained? One possibility is the existence of a spherical diffusion flame where the oxygen from the air diffuses radially into a low velocity laminar vortex-breakdown. The blue spheres that lopped down the aisle in the Jennison sightings may have been natural gas spherical diffusion flames. Another consideration is that combustion studies with swirl burners have demonstrated that as the Reynolds number is increased the flame becomes more closed especially at the rear of the bubble recirculation region where fluid is moving in and out and there is turbulent asymmetric flow.

"Flaming gyroscopes"

How does spherical ball lightning remain stable when it can apparently collide or bounce off objects? The reason is that a burning vortex is analogous to a flaming gyrating top and should have an inherent stability about its rotation axis. It will not be easily deflected from this axis because of its large angular momentum. This might explain some deflection encounters. "Bouncing" or "bobbing" to an observer might be explained by the vertical movement of vortex-breakdown. The vortex-breakdown would not bounce as such, but would respond to changes in air flow that would move the recirculation bubble up and down from the ground as if it were bouncing or lopping along.

Now I will explain the observations of whirlwinds containing fireballs. Theories that posit an independent ball weighing more than air can explain this observation on the basis of the suspension of ball lightning in the vortical flow like a ping pong ball in a jet of air with the Bernoulli Effect is invoked. But the problem is to prevent rapid cooling as air flows around or through the ball as well as centrifugal separation form the vortex. The more appealing and logical idea is for the fireball to be part of the same air flow distribution as the whirlwind, as in combustion within the vortex-breakdown phenomenon.

Glowing Cylinders

Ball lightning with cylindrical geometry like tubes, and "rockets" are quite frequent in published accounts of ball lightning. For instance, six cases of unidentified luminous cylindrically- shaped objects were described in Corliss (1982). He cited an author who reported an object seen at Lamberville, New Jersey in July 1879. A fiery cylinder was seen which measured about one meter wide by about two to three meters high and had a whirring sound.

Another cylindrical sighting (Corliss 1982) took place around 1907 in Burlington, Vermont. An object shaped like a tornado was seen hovering 15 meters from the top of some buildings. It was estimated to be 2 meters in length and around 20 centimeters in diameter. Fiery projections emanated from points on the flaming coppery-red surface. As the object started to move, breaks in the flame surface became apparent, with flames protruding out. There is a strong suggestion of combustion process taking place in this sighting as being similar to other reports of luminous activity in tornadoes. The flames emanating from the cylinder clearly indicates a combustion process, while the rupturing is explained as changes in air flow with extinguishing of flames, which incidentally is an observation not easily understood in stable plasma models of ball lightning. This Vermont sighting is remarkably similar to the "billowing flames" of the UFO seen at Waikawa (see part II of this book).

Hence the glowing cylinder is explained as a shape consistent with a flaming vortex funnel.

Spiraling air flow pattern at Crail, Scotland

Ball Lightning at Crail

Figure 15: Horizontal vortex- air spiraling into the funnel

The next observation showed a remarkably close association between a whirlwind and ball lightning. The spiraling air flow pattern indicated that this burning whirlwind and ball lightning could be one and the same phenomenon. Campbell (1982) reported a ball lightning object at Roome Bay, Crail, in Scotland in August 1968. He compiled evidence from four people, three of whom actually saw the ball lightning. The sighting took place on a fine day, although a postman noted that the noon sky had a hazy copper color. Elizabeth Radcliffe at the west-end of the beach saw a shimmering orb-shaped object approaching her with luminous trails of light extending into the main glowing region. The diameter of the object was 200 millimeters and 0.5 meters off the ground and appeared to be rolling forwards to the beach on a horizontal rotation axis. As the rolling object moved it became more concentrated in its form with increased light intensity.

Another observer Kathleen Cox also saw the ball lightning object revolving clockwise on a horizontal axis, and "rolling up". My assessment of this case is that the orientation of the vortex was along a horizontal plane and the surrounding air flow spiraled inwards to the vortex core. The Weather article was illustrated with a distinctive spiral pattern indicative of a vortical flow pattern. I have seen such spiral air flows within the 0.7 meter diameter vortex generator. The luminosity was thought to be equivalent to a 100W tungsten lamp. She later heard a loud explosion which is consistent with a gas explosion brought about by a sudden reduction in the fuel-to-air ratio bringing it within the explosive regime. The vortex burner theory predicts that ball lightning should end its life, either with a whimper, or with a bang depending on this fuel-to-air ratio. If the concentration of the gas left in vortex-breakdown is within the explosive regime then the ball will explode, but if outside this limit the ball would end with a whimper.

There are additional clues to suggest that this Crail ball lightning sighting fits the vortex burner hypothesis. The weather conditions at the time were particularly suitable for atmospheric vortex genesis along the beach front. In fact a whirlwind threw deck chair in the air. The reported hazy atmosphere maybe have been a combustible gas cloud. Such "haze" has been seen prior to an earthquake and could be natural gas. Eyewitness accounts of luminous spiraling "lines" of light feeding into the main fireball may well be threads of methane gas already burning as a diffusion flame as it enters the main combustion region of the vortex. Such luminous regions are not easily explained in many plasma theories.

SYSTEMS OF FIREBALLS.

Another puzzling feature of ball lightning is its ability to divide into two or more fireballs forming a more complex system. Amazingly, these fireballs can repeatedly split and recombine again without annihilation. To explain such ball lightning sightings I propose to use two main methods by which a vortex can divide into a number of smaller vortex-breakdowns, which I will now discuss.

Vortex splitting

The first method is called "vortex splitting", whereby the parent vortex can transform into subsidiary vortices rotating about a common center. Vortex splitting has been demonstrated experimentally in a "Ward" vortex generator, named after the scientist who first used it. The phenomenon has also been observed in tornado systems. Vortex splitting involves a set of vortices which are able to typically translate through the air while grouped together. Provided each of the vortices is in a state of vortex-breakdown there seems no reason why each vortex could not burn a gaseous fuel. In such a situation there is an orbiting system of flaming balls. When there is a decrease in the swirl angle, as observed in a vortex chamber the daughter vortices recombine giving the appearance of merging to form a single fireball.

This vortex splitting phenomenon is consistent with several ball lightning sightings, including the following ball lightning sighting cited in the work of Singer (1971). The case involved a couple who were scaling a mountain in Switzerland when they saw yellow-colored fireballs (carbon radiation emission?) in a roughly horizontal formation along a ridge in pouring rain and snow. These fireballs combined to form a large ball and then began to split to form red and blue globes. This strange antic was repeated a number of times, over a time span of several minutes.

Multiple vortex-breakdown.

The second property of vortices, called "multiple vortex-breakdown" takes place when two or more vortex-breakdowns originate from the same vortex and are usually vertically stacked one on top of the other. Horizontal multiple vortex-breakdown also seems possible for a horizontally-inclined vortex. I have seen this phenomenon in the laboratory Studies (see photograph). In any case the individual breakdown regions are still connected with the main air flow of the vortex as it moves upwards.

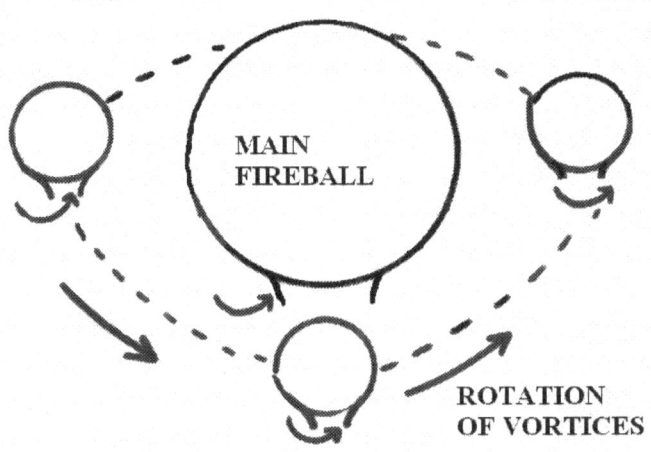

Figure 16 : Vortex splitting - fireballs rotating about a common center

This phenomenon would explain the observation of two or more ball lightning globes on top of the other with a "luminous thread" linking them. For instance, Zou (1988), from the Institute of Atmospheric Physics in Beijing, reported on strange fireballs seen in fair weather conditions.

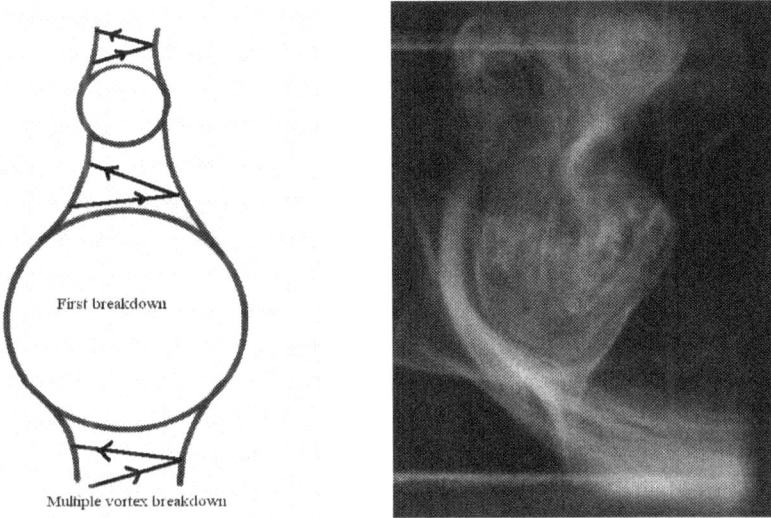

Figure 17: Experimental *multiple vortex-breakdown* (Coleman, 1990)

The event was subsequently reported as a ball lightning "UFO" observation to the 1st International Conference on Ball Lightning, held at Waseda University, Japan. The sightings were observed from an aircraft of the Xingjiang Aviation Company of CAAC. The fireball was spherical and emitted a white beam which was said to be like a search light. The fireball then broke into two parts, one ball above, and a bean-shaped object below. High speed rotation was noted and eventually the lights combined to form a ring of green light. The ring light decreased in size and then finally extinguished.

Multiple vortex-breakdown provides a good working explanation of the above Chinese sighting because of the relative position of the objects-one above the other. The observation of "spinning at high speed" is additional evidence pointing to the existence of a strong vortex.

Although vertical multiple vortex-breakdown has been reported experimentally, in vortex chambers, (eg Coleman, 1990) I am unaware of any published sightings positively confirming its existence in naturally occurring vortices.

MOTION.

A variety of motion has been attributed to ball lightning, including: zig zag, bobbing up and down at ground level, lopping along the ground, rolling along the ground, a bouncing type of motion, and movement with or against the prevailing wind conditions. Each type of motion can be satisfactorily explained by the component motions of a vortex and vortex-breakdown. For instance a vortex may be moving in a downward arc to the earth while at the same time vortex-breakdown is moving up and down along the vortex column. Vortices have been known to move in a zig zag motion. Dust devils have the ability to move with or against the prevailing wind conditions. This could account for ball lightning motion against the wind since vortices have this capability.

Many alternative ball lightning theories are unable to account for such motion, especially a self-contained ball of plasma, without a mechanism to propel itself along. There may be some models that incorporate such a mechanism but they appear to be fewer. Such hypothetical balls lack any obvious means of propulsion to move against the wind, or to remain stationary in high speed air flows over the wing tips of aircraft. Such fireballs would be swept along with air stream.

The bobbing up and down or lopping motion could be explained by the characteristic behavior of vortex-breakdown seen in vortex chambers with very slight changes in air flow. I have seen this phenomenon in my vortex chamber where vortex-breakdown rises and falls along the vortex generator's vertical axis of symmetry depending on the decrease or increase in air flow to the vortex-breakdown zone.

Counter motion of ball lightning.

An unusual ball lightning motion has been reported where ball lightning moves away from an approaching observer. Alternatively, if the observer moves away from the ball lightning it starts to follow. The event below illustrates this peculiar movement. Ashpole (1995) described an incident where a strange transparent ball of light, almost as if it was intelligent, reacted in this "counter motion" way to walking movements.

Counter motion has been observed in connection with UFOs. The same explanation is applicable to UFOs too. This motion can be explained on the basis of changes in the volumetric air flow rate into the UFO vortex. If the flow rate is reduced the vortex-breakdown region shifts upwards. But when the flow rate is increased, the breakdown zone sinks downwards. This movement response is consistent with other vortex-breakdown studies. Now to apply this to a person approaches a hovering vortex, with say, air flow spiraling inward. The person acts effectively as a solid obstacle to decrease the rate of air flow into the vortex, sending the breakdown zone upwards and away from the observer. Conversely, if the observer moves away, the air flow increases because the person is moving, progressively, further and further away from the vortex air stream. This odd movement may give the impression that the object is able to respond intelligently to observer movements.

Following vehicles and aircraft

Another type of motion behavior is ball lightning's ability to following alongside, on top, in front or behind vehicles. What is actually happening is that the vehicle is either creating a vortex air flow behind the vehicle or the basic vortical air flow system has attached on to the car with the car's motion maintaining the integrity of this vortex. Provided the vehicle is moving through an atmosphere of natural gas, the fireball would continue to exist for long lifetimes, several hours in some cases.

There are a handful of reported cases of ball lightning moving with vehicles. Dr. Nickel, a German gas dynamicist in 1988, described an unusual case near Berlin, where two witnesses in a car watched lightning strike a car in front (Nickel, 1988). This lightning stroke apparently slid down the back of the car and curled up to form a one meter diameter ball. This ball then proceeded to move with a velocity of 4 ms⁻¹ towards the car and apparently was run over by the car. A smell of burning rubber was noted. The ball then quietly disintegrated about 300 m behind the car. To an observer the perceived effect was fireball creation from a lightning strike which transformed into a plasma ball. In actuality the lightning strike may have created a burning vortex that hovered around the car tires.

In a contribution to a book entitled *"Lightning"* edited by R.H.Golde and published in London by Academic Press, Stanley Singer described a report of ball lightning on the wing tip of Boeing 727 aircraft. Wing tips are known to produce vortex-breakdown. Such events more than likely explain the famous foo fighter phenomenon of World War II (see Part II)

ROTATION.

Observations of rotating ball lightning are quite common and provide a valuable source of evidence supporting the vortex burner hypothesis. Charman (1982) said that a sizeable portion of reports involve ball lightning as either rotating or rolling.

Although several authors on ball lightning have described rotating flames and balls of fire, to my surprise, I could find no reference to rotation in the earlier review of Barry (1967). On the other hand, (Cade and Davis, 1967) cited the ball lightning review of Dewar, who found 46 cases out of 513 in which rotation was seen. There are many individual cases of rotating fireballs sprinkled throughout the scientific literature. The characteristics of the Norwegian ball lightning fireballs, "rullende lyn" show visible signs of a vortex. Rullende lyn commonly rolls along the ground, wobbles or precesses like a top, and in Sweden they have been known to pull up lamp posts from the ground. Aerogel phenomenon certainly cannot uproot a lamp post like a strong vortex.

Observations of a rotating flame.

Ball lightning has been observed first hand as a rotating flame. Such observations, of rotating flames in connection with ball lightning are further evidence supporting the vortex burner hypothesis. These reports provide a body of circumstantial evidence that point to the vortex hypothesis as being a correct interpretation of the phenomenon. Here are some samples of published ball lightning observations indicating a rotating flame or fireball. I quote directly from my 1988 *Kugelblitz* report (Coleman, 1990).

"After a severe lightning strike in a storm, a yellow spherical flame, 12 cm. in diameter, and spinning like a top appeared in a house. A strong sulphurous odor was noted. The flame did not explode. A cook watched the ball appear from a mere distance of one meter. She then ran away" (Wood 1930).

That the flame was *"spinning like a top"* strongly indicates the presence of a vortex.

The following sighting records a rotating spiral structure adding to the evidence supporting the existence of a spinning vortex,

"And in a plane in 1948 a witness recorded seeing a ball of fire, orange-yellow in color, slightly larger that a tennis ball rise from under the cabin. a dark grey-violet layer surrounded the bright center. the ball was 2-3 cm. thick with a sort of tail, as if it were a rotating spiral... (Baratoux 1952).

Further evidence alluding to rotating ball lightning comes from Ofuruton et al (1997) who found that 39.80% of the 250 ball lightning cases they catalogued in Japan were of a sphere with a tail.

Another case of visible rotation was reported by Chaggar (1982) on a woman living in Nairobi who suddenly awoke to see a burning disc the size of a plate hovering around the door. The disc sparkled around the edges and began to contract to a small ball. Slow rotation was noted. The central region was blue and heat was detected. Following the disintegration of the ball, a smoke cloud was left, which did not readily diffuse, but moved around the room and into the kitchen. It then disappeared. Some mechanism contained the smoke while it made its way into the kitchen. My guess is that the vortex ceased being in a combusting state and the vortex-breakdown still remained but moved into the kitchen. The smoke was retained in the recirculation region of the vortex-breakdown region and the vortex subsequently decayed. This non-dispersed smoke ball observation seems to neatly fit the category of so called "black" ball lightning.

If rotation is such an inherent property of ball lightning, why is it not always reported? I have some possible reasons for this. To see rotation in a non-combusting vortex, as I have discovered with vortices in my vortex chamber there must be some visual cue to indicate the revolving air mass. If there are no tracers in the air flow the detection of rotation may be difficult. A more turbulent flame might mask the effects of rotation as the vortex remains invisible. Smoke or small particles, when lifted into the vortex could indicate rotation. If a smooth or laminar flame is present better signs of rotation. Signs of rotation may also be overlooked because of other dominant features occupying the attention of the observer, such as color or brightness. The spin rate may be too great for the eye to detect rotation in the flame. Near the town of Tom Price, in the Pilbara region of Western Australia the following unusual sighting was attributed to ball lightning but it could equally have been classified as a UFO:

*"Our original barbecue observers, being some 200 meters directly below it by now, reported that **it was an intense spherical ball of orange-red fire with the fire swirling in a spiral pattern and the flames disappearing internally upwards into a central black "hole" or void within the***

spherical mass of flames. The fireball had no tail and made no noise at all - there was no ground seismic wave as experienced in many other recent Australian fireball events. It was described as a sort of "implosion ball of flames" with all the fire or flames originating in local space outside the fiery sphere-like form, the flames being sucked into the center where they disappeared - "like a moving plasma ball in a local space-time warp around a central black hole" - "Never ever seen anything like it before - therefore difficult to describe accurately"... ".. (Dr B. Mason).

This is an impressive description. Many vortices only show circumstantial evidence for combustion in a vortex. This observation actually shows a 'fire swirling in a spiral pattern' strongly suggestive of rotation inside a combusting vortex-breakdown. The ball of flames is also part of the expected description of such a turbulent flame. The flames being 'sucked into the center' is also a feature explained within the vortex-fire ball theory. Such a swirling flame phenomenon has also been reported in the Waikawa Bay UFO episode that I will describe later. Such accounts are
rare but extremely important since they indicate unequivocally the existence of the meteorological phenomenon I have been talking about.

An explanation of the dark zone
The dark hole in the middle when the fireball was seen at night is indicative of a burning vortex-breakdown. This effect has been seen in connection with another ball lightning event seen in Tennessee. The object was described by an observer as pear-shaped and orange-yellow. The distinctive feature was that the object was divided into two halves along the vertical axis by presumably a flameless cylinder. If the theory is correct there is a need to interpret this. I suspect that the hollow cylinder is actually the funnel of the vortex connected to the central reverse flow zone of air of the vortex-breakdown zone and a funnel region of the vortex above. It could also be just the reverse flow zone if the funnel is not visible. Such a feature was undetected in the smaller burning laboratory vortices but evidently a feature in the larger vortex fireballs seen naturally in the atmosphere.

BALL LIGHTNING SEEN IN FAIRWEATHER
One myth surrounding ball lightning is that it is always directly connected with a lightning strike. There is now direct evidence against this view. It is now acknowledged by several ball lightning researchers that a large number of ball lightning events are not associated with lightning strikes at all. Ohtsuki and Ofuruton (1988) documented 2060 cases of ball lightning where 550 were seen in fine clear weather or in cloudy conditions. Grigorjev et al (1988) reported 203 sightings in clear weather or non-

thunderstorm conditions.

Fine weather ball lightning poses a difficult obstacle to lightning-based ball theories. One argument in defense of these theories is that ball lightning could be produced by fair weather lightning strokes, which do happen. But in that case there would also be the sound of thunder which is not always reported in these fair-weather sightings. Some may suppose that the ball an observer sees in fair weather was in fact previously produced in a thunderstorm. However it is unlikely that all fair-weather balls can be explained in this way since ball lightning can be produced in fair-weather not directly connected with lightning by storms.

BALL LIGHTNING IN ENCLOSED SPACES.

Eyewitnesses report fireballs seen inside buildings and in airplanes. There are three methods whereby ball lightning may obtain access to an enclosed space. First, the ball lightning in question could have entered through some opening in the form of a mesh or crack in a wall or window. The other option is that it may have physically forced its way in. The other method of entry is that a hot combustion flame could have melted a section of a window. Finally the fireball may have actually materialized within the enclosed space. Clearly the vortex fireball theory can apply to the first two situations. The third method of ball lightning entry appears to require some external source, such as a radiation beam focused in the direction of the enclosure or an electric field. With the beam idea a plasma could appear to pass through a window from the outside. A ball lightning theory invoking some kind of microwave or maser theory might seem appropriate. In a completely sealed enclosure, a vortex could not suddenly appear inside when it was previously on the outside, unless there was a combustible gas present and there was a shock wave induced vortex. Radiation beams cannot be entirely ruled out. However to my knowledge there is no scientific evidence suggesting that such radiation emission has sufficient intensity to produce a plasma inside a room. The Russians carried out this work around the hills of Moscow.

Materialization of ball lightning at windows and later penetration through closed windows, without any effect, are very rare. In fact, I could only find one such report in Powell and Finklestein (1970), which is also cited by Corliss (1982). If this one observation was an authentic interpretation then some other phenomenon is involved. But many properties of ball lightning can be explained on the basis of vortex burner theory. This is only one case and in contrast there are far more cases where ball lightning actually entered through open windows, through metal grills, and in some situations actually penetrated the window leaving a hole in the glass. Corliss (1982) cited a report on such penetration where a fireball with a fuzzy boundary came in through a wooden screen and melted a hole in a glass window 28cm in diameter. The ball appeared to have internal motion.

Ball lighting inside an aircraft.

One challenging observation is to explain ball lightning appearing inside an aircraft using the vortex fireball theory. A metallic fuselage would provide an almost totally enclosed environment and act as a Faraday cage to prevent an external electric field developing in the interior of the fuselage.

Jennison (1969) described a remarkable incident in which a deep blue spherical ball lightning moved at 75 centimeters above the floor of an aircraft straight down the aisle at 1.5 ms-[1] between startled passengers. It was about 20 centimeters in diameter judging from the size of surrounding objects. It was not possible to say whether or not it was spinning. It cannot be ruled out that the plane could have traveled through a methane gas plume. Gold has described evidence of such lighter-than-air (slightly greater than half that of air) gaseous plumes that can reach 50 000 ft and 200 miles in diameter. Furthermore methane gas has been observed in a near aircraft disaster while flying over an Indonesian volcano.

Generating the initial fireball.

The Jennison eyewitness account is difficult to reconcile with the burning vortex concept. It may be that this luminous entity was a low energy passive discharge phenomenon. How does a vortex manage to get on the inside of an aircraft? A vortex certainly could not move up to a smooth section of plane fuselage and force its way in. It may be able to "squeeze" its way through a gap such as in an aircraft door. This seems highly unlikely in modern aircraft. Only one possibility comes to mind if the vortex burner is applied. A lightning stroke would need to impact on the fuselage of the aircraft to cause a small section of fuselage metal to disturb the air inside to create a vortex. Almost simultaneously there could be ignition of the natural gas by a shock wave to the natural gas or methane/air mix. Shock ignition of such gaseous fuels is a well recognized method in combustion science. Shepard in 1949 demonstrated that methane shocked by adiabatic compression can ignite the methane with oxygen at temperatures as low as 120 degrees Celsius. (Gaydon and Wolfhard, 1953). Electrical ignition seems a more remote mechanism given that the fuselage is almost a hollow metal cylinder and electric currents would disperse over the surface of the fuselage.

There is support for the lightning strike shockwave scenario. In the Jennison sighting there was a lightning strike, with only a time lag of a few seconds before the ball was seen. A lightning stroke hitting the outside of the front cabin directly could set up vibrations by a shock wave in that section of the fuselage and thereby generate air vibrations adjacent to, but on the inside, producing a small vortex.

How does the fuel gas get in?

Next there is the problem of how the natural gas worked its way into the plane's interior. The possibility exists for air mixed with methane gas to enter the cabin at ground level through a series of air ducts and leave through a couple of exit valves in the belly of the aircraft fuselage. In some planes the air ducts are connected to the outside of the plane, to the air compressor in the engine area at the front of the plane. There is likely to be a weak overall air flow through ducts originating from the front engine along the center aisle and towards the back of the aircraft. This may help to explain the motion of the ball lightning towards the back of the plane.

The possibility that gas or fine particulates can enter through the ducts is not without precedent. Wendy Tootell (Tootell, 1985) wrote of her ordeal when a British Boeing 747 flew across Mt Galunggung, Java, Indonesia. It was reported that a smoke-like substance came through the planes ducts. On page 22 of the book Captain Moody observed smoke issuing from the ducts located on the cabin floor. There were fires on the wings and coming from the engines. Unexplained revolving fireballs were seen in the jet engine intakes. They were said to be like giant cannon balls, and red spheres. My guess is that these fireballs were created by natural combustion of natural gas in the vortices inside the jet intakes. The plane flew over Java when Mt Galunggung was erupting. Java is part of a sweeping arc of fault lines which is well known for gas fields and mud volcanoes that can emit gas. This gas may have been present at the time of this eruption and burnt in and around the aircraft especially the jet engine intake. All the luminosity in this event was put down to St Elmo's fire but I feel it is hard pushed to explain the revolving fireballs seen in the engine intakes. The intakes lights shone like torch lights. This report is therefore reminiscent of the search light description of ball lighting events and the Ande's effect.

The idea that a flammable gas fuels ball lightning inside an aircraft could be tested by employing a gas detector on board an aircraft. The intention would be to establish a direct correlation between ball lightning and the presence of the fuel gas. The plane would need to have traveled through a region in the atmosphere known to have natural gas, during a thunderstorm with a good chance of lightning hitting the plane. Positive confirmation of this fuel gas by a gas detector would provide evidence for the vortex burner theory and help to explain the Jennison sightings or similar sightings on board an aircraft.

Ball lightning inside buildings.

To account for ball lightning inside buildings, the vortex could be created by a lightning stroke penetrating into the interior of the room. Another possibility is that a small vortex entering a room could be ignited by electrical sparking from a corona discharge, in the form of St Elmo's fire. Such a transient vortex phenomenon would not last long because the vortex would be weak and decay quickly. This may explain the flame-like ball lightning that suddenly appears in a room and then just as quickly disappears. A possible example illustrating this was described in Powell and Finklestein (1970). Apparently a bright yellow ball spontaneously appeared on a tile floor in a kitchen. One second after, the ball exploded with a very bright light and the room had an odor of burnt powder.

INTERNAL STRUCTURE
Holes, protuberances etc in the fireball

Observations of ball lightning show that the surface of the sphere is not always smooth or laminar. Sometimes people have seen glowing blobs swirling around on the inside of the ball. In other reports the ball develops a hole in the bottom so one can see into its interior. Somehow the fireball still manages to maintain its overall integrity. This last observation of asymmetry (lack of symmetry) is a rather difficult problem for plasma theories of ball lightning that requires a strict spherical or cylindrical symmetry in their mathematical treatment. In addition, plasma theories apparently do not have an adequate explanation of observed internal motion of isolated luminous regions or large particulate material swirling within the interior. I am not sure how the field of dusty plasmas can accommodate this.

Then there are observations where the ball suddenly develops glowing tentacles ejecting large material objects from within the interior of the ball lightning. Cade and Davis (1969), page 63, cited a ball lightning in December 31, 1924 at Aberystwyth, Cardiganshire which emitted fiery lumps of material and had luminous protuberances. My explanation is that the fiery matter is centrifugally thrown out of the vortex. The protuberance was most likely a local eddy-like disturbance of air flow which alters the flame structure so that the flame front is "pushed" out. How do plasma theories of ball lightning which require certain symmetries, address this observation? What force will hold this burning material in the plasma?

Some observers report a fibrous or hessian surface while others describe the surface as wriggling, twisting bands of light. Corliss (1982) quoted the account of a man walking during a thunderstorm who saw a fireball with the diameter of a tennis ball with fibers or strips of light moving rapidly in a type of rolling motion. In another case someone saw a ball about two thirds of meter in length. The color of this sphere was a blue-green color and consisted of wriggling "spaghetti" of light with each being a few millimeters in diameter.

This hessian appearance may be just another description of what other observers saw as the filaments of light moving around. What these observers may be seeing is the turbulent flow of the spherical flame rather than a laminar flame. Laminar flames at lower air speeds, will transform, at high speeds, to quasi-spherical flames with a "boiling" turbulent interior. Obviously this latter type of ball lightning would possess a larger kinetic energy ball because of the higher air flow speeds.

A less likely explanation for the woven cloth appearance might arise from an internal assemblage of fibers captured by the vortex and retained in the swirling recirculation region of the vortex-breakdown region. These fibers manufactured in the hot interior of the burning vortex by the melting could be subsequently drawn out into fibers-a process not unlike the manufacture of candy floss form a revolving candy floss machine. I have not come across any evidence of such processes in connection with ball lightning although there are cases of UFOs ejecting fine fibrous hair-like material.

It is conceivable that lightning, once it had produced a vortex, could simultaneously dislodge and heat up matter to form a fibrous mass. Furthermore it would not be unreasonable to expect electric charge to be transferred to the fibrous assembly and trapped in the recirculation of vortex-breakdown for several seconds which is then swept up the vortex. I demonstrated experimentally that coarse fibrous clumps can be held in circular orbits of the recirculation vortex-breakdown region in a 0.7 meter diameter vortex chamber with residence times in the order of several minutes (Coleman, 1990). This electrically-charged fibrous assembly would account for electrostatic attraction to objects in a room which has been reported.

WHERE IS BALL LIGHTNING CREATED?

Ball lightning has been seen in many different locations: in connection with volcanoes, regions which are prone to thunderstorms, on flat lands, on mountain ridges, out at sea, traveling along the side planes or on their wing tips, inside planes and buildings. Collectively these reports suggest that the conditions required for ball lightning formation in this type of location are, in fact, quite common. The vortex fireball theory does not require land with highly unusual and specific characteristics, such as areas of nuclear activity, or flat land areas as required in the theory by Peter Handel. There are only four ingredients to create ball lightning.

Four requirements for ball lightning genesis.

The general principle behind the vortex burner hypothesis is that it does not require any special process that is not already ubiquitous in nature. There is no need to posit antimatter, black holes for ball lightning, or the like. What is required is the coincidence of an ignition source (eg. fire, lightning or small electrostatic spark), and oxidant (almost always oxygen in the air) a source of fuel gas or combustible (eg methane gas, wood) and a vortex. The vortex and perhaps the ignition source, in principle, could be found just about anywhere in the lower atmosphere and on the Earth's surface. Even the vortex such as the tornado could produce electrical discharges to re-ignite the gas circulating in the air. On the other hand the requirement of fuel gas such as natural gas required by the theory narrows the range of places where ball lightning genesis could take place.

Natural gas (mostly methane) would be the most commonly occurring fuel gas to power ball lightning but the possibility exists for other fuels and oxidants to react inside the vortex. Hydrogen sulfide is a well recognized gas that is spewed out in volcanic eruptions and lurks around swamps and mines. Hydrogen has also been detected during volcanic eruptions and also during earthquakes (Gold, 1987). Hydrogen-fuelled vortex fireballs may be a possibility in nature. However the density of hydrogen (0.08 gcm^{-1}, at zero degrees and one atmosphere pressure) is about ten times lighter than methane. Because of its greater buoyancy it may be very difficult for the fuel gas to be premixed with air and enter a vortex. On the other hand, there is the possibility that very fine dust could also combust inside vortex-breakdown. Dust has been known to burn like a flame (Gaydon and Wolfhard, 1953). Fireballs in some dust storms may therefore be a possibility.

There are basically two types of region where natural gas would be found naturally. One area would be swampy locations where bacteria generate marsh gas and/or hydrogen sulfide. There have been increases in atmospheric concentrations of methane around swampy areas (Singer, 1971) cited this work). The second type of location is at natural gas outlets from the Earth's crust, especially from volcanic fissures and vents along active crustal plate boundaries and their intersection. These are the major regions where ball lightning should be observed.

Ball lightning-Bay of Islands, NZ

I came across a rather intriguing observation which is possibly the first New Zealand ball lightning field observation in the scientific literature. The sighting was described in a Letter to *Nature* by Professor Burbridge and D.J. Robertson (Burbridge and Robertson, 1980).

The sighting took place at a house situated near a River mouth of Te Ngaere, north of Russell, in the Bay of Islands, along the same coastline.

The witness to the event was Mrs E.V. Sale who saw a metallic and milky colored light which came in under the door after lightning struck near by. The light then proceeded to hover over some stationary tools. While suspended over the tools it pushed out "oily tentacles" which then retracted. Detailed magnetometer measurements were done at the house where the ball lightning was seen. The object's motion kept more or less along the peak regions of high magnetic field. However there was no conclusive evidence to say that the glowing form was actually influenced by the magnetic field.

What mechanism produced the "tentacles"? Changes in air flow could alter the shape of the vortex-breakdown influenced by the surrounding topography. Movement of charged particulates of the flame may be attracted to the tools thereby affecting the air flow and flame envelope.

The vortex was probably weak, since it was carried under the door, possibly by a draught of air. A strong vortex on the other hand, would be able to move around the room, and against any prevailing draughts.

Features of this case warrant further consideration in the light of the vortex burner hypothesis. Mrs Sale's house was situated at the mouth of a swampy river estuary. It is not unreasonable that at the time of the storm there may have been a build up of marsh gas or possibly hydrogen sulfide. Marsh gas, composed mostly of methane may have originated from the anaerobic bacterial breakdown of vegetation in this swampy environment. A second option is that the origin of the suspected natural gas could have come from the ground. Field work would be required to investigate this possibility. Geological records of the area might reveal the presence of natural gas, while detectors could be set up in the area to monitor the methane variations near ground level.

The color of the ball lightning that came under the door was a bluish-silver color is noteworthy. This color is partly consistent with the combustion of methane which can burn with a blue color. The absence of any odors or sounds in the original report is consistent with methane combustion. There was no mention of odors or sounds although this could also mean that these features were simply not included in the final report.

Criticism of natural gas ball lightning theories.

Critics of hydrocarbon theories for ball lightning have pointed out that the natural abundance of methane is insufficient to be able to explain ball lightning sightings which seen in areas deficient in marsh gas. This criticism is easily dismissed because there are many regions that are connected with natural gas. As far as the vortex burner hypothesis is concerned there only needs to be a flammable gas and natural gas is only one among a few candidate gases, like hydrogen sulfide. In addition Gold

(1987) pointed out that unaccounted increases in methane concentration in the atmosphere have been recorded although not been adequately explained. In a short article Gold on website at the following url: (people.cornell.edu/pages/tg21/hazard.html) also listed further evidence of eruptions of natural gas. The pockmarks on the North Sea floor from gas emissions and the anomalous "mystery clouds" inferred to be methane or hydrogen that formed large plumes even up to 50 000 ft. Several pilots flying from Tokyo to Alaska saw this cloud on April 9th, 1984. Gold also mentioned in the same article that slow moving flames and fireballs have been seen over areas having a history of gas emission from sand beaches and from the sea.

Natural gas emission during earthquakes.

Is there any evidence in the literature in support of natural gas emission during earthquake activity, or along fault lines to act as the fuel gas for ball lightning? Gold (1987) provided several observations as evidence to substantiate this claim. He collected these observations to support his hypothesis that the Earth is a vast abiotic (not arising from plant and animal remains) reservoir of natural gas and oil. There are critics of Gold's hypothesis regarding the origin of the earth's hydrocarbons, like coal and methane and I am aware that this is controversial. He has had some failures and successes regarding hypotheses. The neutron star hypothesis was correct but he did not convince others of the steady state theory and the prediction of a thick covering of dust on the moon. Whether or not his deep earth theory is correct is a moot point but there are still several scientific references that suggest evidence of natural gas emission along tectonic faults. Gold believed that gas emission along earthquake zones is widespread and that in some places it is difficult to detect because the gas sources are not as active as in other places. The standard view is that methane is entirely derived from sedimentary rocks and that earthquakes merely disturb this. This is in contrast to Gold's deep earth origin of methane which originated from the Earth's original creation.

Gold was not alone in believing in a greater abundance of methane than commonly imagined. The work of Judd et al (1997) on atmospheric seepage of natural gas on the UK Continental Shelf would appear to lend support to Gold's hypothesis. They contend that the earth's geological source of atmospheric methane is greater than is widely thought. The unabated discoveries of new gas fields in areas not normally considered gas-bearing is in line with this type of thinking.

Gold (1987) gave circumstantial evidence indicating gas emission during earthquakes including, "sand blows", which are conical sand craters seen during earthquakes such as the major New Madrid earthquake. Furthermore, pockmarks from a few meters up to 200 meters in diameter have been seen on the seafloor coinciding with the gas and oil fields of the North Sea.

The sightings of unusual clouds, which looked like gas hanging above a faulting zone, have been reported just before the onset of earthquakes. Deng et al (1980) reported a precursory fog, before an earthquake in China, which remained at the location for some hours before the fault slipped. Haywood, in 1823, noted that during the New Madrid earthquake of January 1811 to February 1812 a dark cloud appeared before the earthquake. The cloud was not smoke or an ordinary cloud but similar in some respects to both. This could have been a natural gas cloud fed from some ground source.

If such clouds of natural gas streaming from fissures are quite common, they are easily ignited from electrostatic spark generation, such as the piezoelectric effect from frictional slip along a fault line. That the observations of gas clouds have been seen before a tremor would lend support to a combustion origin for earthquake light.

There is more evidence for a combustion origin of earthquake lights. Flames seen in connection with earthquakes are common. The flames are strikingly similar to some episodes of earthquake lights and the so-called "Andes Lights" effect which I will discuss later. The following eyewitness accounts point to the combustion of a gas coming from within the earth.

There is some evidence that prior to the 1906 San Francisco earthquake there were blue flames moving across the ground and along the crests of the foothills west of the city. Schmidt and Mack (1913) reported a set of gas-like flames from the street. These flames were 8 to 12 centimeters high. The phenomenon was likened to alcohol poured on the ground. Flames were also seen in a meadow and these were 80 centimeters in height.

Further evidence of a combustion process is revealed in the Quebec earthquake of 5 February 1663. Flames were seen and burning "brands" were observed flying around houses (Lalemant, 1663). These aerial "brands" which were presumably flames, would seem to parallel ball lightning behavior. This report demonstrated a clear indication that flame-like entities can move through the air as ball lightning does.

Another sighting reminiscent of the Andes effect took place in the Owen's valley area on 26 March, 1872. Several miners witnessed sheets of flame from the ground. They actually hovered above the ground as well like flaming torches (Inyo Independent, 20 April, 1872).

Further indirect indication of gas emission, cited by Gold, were the occurrence of many dead fish, observed by fishermen near Tokyo, some days before a powerful earthquake rocked the Kanto plain. The dead fish effect was explained by Gold as natural gas diffusion into water which then deprived the fish of oxygen. This natural gas diffusion into the water would account for large numbers of carp that came to the surface of a pool in central Tokyo just a few hours before the main earthquake. The carp were resuscitated when they were put into fresh water. This indicated a lack of oxygen in the water likely to have been caused from water containing natural gas.

Ball lightning seen in natural gas-enriched areas

The vortex-breakdown burner hypothesis predicts a high correlation of ball lightning sightings where there is also a high frequency of the occurrence of vortices, an ignition source and a fuel gas. So although some locations might have a natural abundance of natural gas emission, the area may not have a high frequency of vortex-breakdown production or lightning. Hence, under this theory, ball lightning might be rarer here than at localities with a high incidence of all three factors. So it would not be unusual that many fireballs are seen in some volcanic eruptions. As I have stated earlier, the theory predicts more ball lightning events from muddy volcanoes than other volcanoes. Mud volcanoes not only release methane but high speed gas streams from the volcanic vent can create electrostatic sparks. Swirling air from an eruption may generate numerous vortices under some conditions. In fact whirlwinds have been seen in connection with the well known Surtsey volcanic eruption (Thorarinsson and Vonnegut, 1964).

Conversely the theory might explain the absence of ball lightning in certain areas. For instance, Handel (1997), cited the well known report by the meteorologist, K. Berger, who said that in all the time he had observed ordinary lightning, from the Mt San Salvatore Observatory in Lugano, Switzerland, he had not once seen ball lightning. Perhaps the real reason for ball lightning's absence is that this site is not endowed with flammable gas emission.

Gold noted several areas where natural gas fields can be found, and so too the possibility of ball lightning production. Is it any coincidence that a possible ball lightning sighting by (Lewis, 1988) at Abu Dhabi, in the United Arab Emirates, in Persian Gulf could have been connected with oil and gas fields nearby? I strongly suspect the existence of natural gas in the atmosphere around Abu Dhabi, on 17 February, 1988 when a lightning storm took place a couple saw a horizontal lightning stroke discharge from cloud to cloud. For one or two seconds following the stroke, a "necklace" of 20 balls of light became visible. This phenomenon was also independently witnessed by friends of theirs some distance away.

World-wide natural gas locations

Other promising ball lightning sites should generally coincide with natural gas producing regions. More specifically, ball lightning investigators need to examine whether there have been any reports of fireballs from such world-wide gas fields as the Tigris valley, and the island arc closely associated with a line of volcanic activity from New Guinea through Java and Sumatra. This island arc pattern also extends into Sumatra into Burma and Southern China. Gold speculated that the Marianas and South Sandwich Islands have the same kind of pattern as other island arcs with volcanic activity and gas production.

Lake Kiva is a rift valley lake which has record concentrations of methane. Other African lakes, like Lake Tanganyika and Lake Baikal, have increased amounts of methane in their waters, while another rift valley in Russia is also associated with natural gas and oil production around its shores. It would be interesting to see if these areas have a history of fireball sightings. The Caribbean Islands have volcanoes and earthquakes and great quantities of oil and gas. The Japanese Island of Hokkaido is rich in natural gas with vigorous gas emission in the Kurile Islands to the north. The Hokkaido coal mines have very high levels of methane. In Japan there are large areas of porous volcanic rocks called "green tuft" with widespread commercial natural gas production.

In Sweden, despite the large areas of granite, which are not normally thought as unlikely to be natural gas bearing they can be reservoirs of natural gas. For instance, the "Sitja Ring", considered to be Europe's biggest crater from a meteorite impact, produces natural gas while Swedish farmers have observed their water wells producing combustible gases.

The East Pacific rise is a gigantic rift stretching across the Pacific with methane fissures along its length. Several natural gas areas exist in the United States. The Anadarko Basin Oklahoma, the San Juan Basin of New Mexico, and the Hugoton-Panhandle fields of Kansas and Texas. Several instances of mystery lights have been recorded in these areas, though not identified as ball lightning. The North Sea Trench is a well known commercial source of natural gas.

Two other types of sites where methane is found could be important in ball lightning production. These are methane hydrates and mud volcanoes. Methane hydrates are frozen mixtures of water and methane and are found in many places in the world, including the Arctic Ocean floor, and the Pacific, where thousands of samples have been found. These formations cap the release of methane and prevent it from escaping - unless of course these hydrates melt.

The second type of location for combustible gas emission is the mud volcano. These volcanoes are said to be generally rich in methane production. According to Gold (1987) mud volcanoes, as a rule, give off mostly methane, while on the other hand lava volcanoes tend to produce a large portion of water vapor and carbon dioxide. Therefore there should be a congregation of UFO sightings dotted along areas where mud volcanoes are located.

More than half of all mud volcanoes in the world are located adjacent to or near Eastern Azerbaijan. One prime example of a mud volcano is Mt Baku near the Caspian Sea, where flames have jumped up to a height of 2 kilometers and stayed burning for eight hours at a reduced height. Therefore the theory predicts that these UFO fireballs should be seen in the Caspian Sea area. This area is apparently rich in sightings from the anecdotal evidence available from websites. The Bakusun newspaper in August 2003 reported on UFOs flying over an oil facility. Even a solar astronomer from Shamakha Observatory, Rovshan Salmanzadeh saw a UFO in 1999. A Russian Academician, Fuad Gasimov, head of the Seismological Department of the National Aerospace Agency believes that there are UFO bases in the Caspian Sea.

Frequency of ball lightning greater in the North Island?

A rough prediction follows from the vortex burner theory. The North Island is volcanically-active with rich sources of natural gas. If this is the case, then vortex fireball production may be more frequent in the North Island than in the South Island. It may also have a higher incidence of tornadoes. This correlation may also be applied to world sightings.

Statistical surveys would enable a test of the above prediction but the problem is that there appears to be no significant New Zealand ball lightning scientific database to draw upon to test the validity of such a prediction. The Meteorological Service may have the odd fireball reported in their literature but nothing like a frequency distribution map for New Zealand. On the other hand, countries like Japan have more ball lightning data enabling such a prediction to be tested.

High frequency ball lightning locations

The regions where there is a high incidence of ball lightning occurrence should be linked directly to areas which a high frequency of vortices, natural gas, and a source of ignition. As I have said earlier, these regions are most likely along sections of the Earth's plate boundaries. This work in the ball lightning field, as far as I know, has not been done and it would be a further opportunity to test the vortex burner hypothesis.

Alternative research by some authors in UFO research has shown a high correlation between certain UFO sightings and earthquake activity. These observations, while either dismissed or ignored by many professional scientists are consistent with the vortex burner theory.

Japan has its fair share of ball lightning observations and is no doubt a favored site because of the tectonic activity and natural gas abundance there. For example, Kobe is located near the intersection of three major plates. It would seem reasonable to propose that ball lightning should have been seen during the 1993 Kobe earthquake. Indeed eyewitness accounts obtained from fisherman and hotel guests show that unexplained lights, including fireballs, were witnessed during the Kobe earthquake. These fireballs were reported by ball lightning researcher Professor Ohtsuki (Smirnov, 1996).

FURTHER SCIENTIFIC PUZZLES EXPLAINED

As I researched ball lightning I found a number of other enigmas in the scientific literature that were considered quite separate from the ball lightning problem. The vortex fireball hypothesis looked to me like an attractive way of bringing together these atmospheric anomalies and explaining them with a single unifying theory. There may be more such scientific anomalies in the literature than I have reported here. I briefly report on each of the phenomena and I am optimistic that the vortex explanation will be adopted when scientists come to the realization that it is the best explanation.

Fireballs in volcanoes, tornadoes, and water spouts.

From "The Aerial World," by Dr. G. Hartwig, London, 1886, page 262. [Courtesy NOAA Photograph Library]

I will begin examining these anomalies by taking a closer look at fireball objects seen in volcanoes, tornadoes and water spouts. Singer (1971) reported that fireballs have been seen in association with volcanoes.

From the perspective of the hypothesis a solution immediately becomes apparent. All the necessary ingredients for fireball production, according to the hypothesis, are present during some volcanic eruptions. Certain volcanic eruptions generate not only natural gas emission, but also vortices, and even lightning or electro-static spark from the volcano vent to ignite the gas. For example, two of the necessary ingredients for ball lightning were present during the Surtsey volcanic eruption which was reported by Thorarinsson and Vonnegut (1964). In the eruption both whirlwinds and lightning strokes were reported. The omission of any mention of fireballs suggests that perhaps gas emission was not natural gas but mainly steam and carbon dioxide which lava volcanoes are generally known to possess (Gold, 1987).

Some early authors have associated ball lightning with tornadoes or vortices but these explanations were not universally accepted. Perhaps the theorists may not have addressed the association of a lightning strike with ball lightning. There would also need to be some explanation of the luminosity. Knowledge of vortex-breakdown and its ability to burn a fuel would not have been available to these earlier investigators. Faye suggested in 1890 that ball lightning could be explained as small vortices obtained from whirlwinds or tornadoes. This speculation was based on several fireballs the size of billiard balls seen in a tornado in the same year. The really interesting property of these fireballs was their ability to "bore" 8 centimeter holes in glass windows facing the storm. This observation is revealing because recent observations of ball lightning also demonstrate this property to punch and melt a hole in a window. A burning vortex would appear to be the answer.

Botley in *Weather* described in 1966 how a tornado could "hatch" fireballs from the lower tip of the spout. Bonacina in 1946 reported on the Widecomb disaster in which a church was demolished by a whirlwind. A stylized drawing at the time showed a fireball connected to a funnel.

How do fireballs in tornadoes arise? If fireballs are to be ejected or "vomited" from the main tornado then they are likely to be smaller secondary vortices originating from the same swirling air mass during the eruption. This is then clearly a situation of vortex splitting. In the second scenario a fireball arises from the main tornado itself. The luminous region high up along the vertical axis of the tornado, where the tornado is undergoing vortex-breakdown, should have an extensive amount of combustion there. The following observation is consistent with this last idea. An observer, called Montgomery, described a 100ft thick band of deep blue light 900ft off the ground in the 25th May 1955, Blackwell, Oklahoma tornado (Vonnegut, 1960). Another such "band" or ring-shaped light was also seen in the Silverton, Texas tornado of 1957 (Silberg, 1966). Vaughan and Vonnegut (1976) recounted the observation in the Huntsville, Alabama, of a series of nocturnal tornadoes. One especially spectacular display was dazzling "egg-shaped" yellow light, estimated to be 400-500 meters wide, which strangely turned on for 2 seconds, turned off, and then turned on 5 seconds later. In the same storm ten to fifteen fireballs were visible at one time which moved in arcs like a fireworks display. The various balls of fire reported by Meaden (1989), such as the "huge revolving ball of fire" of the Newbottle tornado of 30 September 1872, would also fall into the category of being a fireball embedded on the main axis. Ball lightning has also been seen in water spouts (Singer, 1971; Meaden, 1989). This is an interesting type of observation. Water spouts are air vortices capable of "sucking" up a quantity of water from the surface of a body of water such as a lake or the sea.

Why doesn't ball lightning quickly cool down like a red hot poker being thrust into cold water? Self-containment theories of ball lightning, like the hot metallic vapor ball lightning theories, and electrical theories, by their nature, would lead to such rapid cool down. In the latter case this

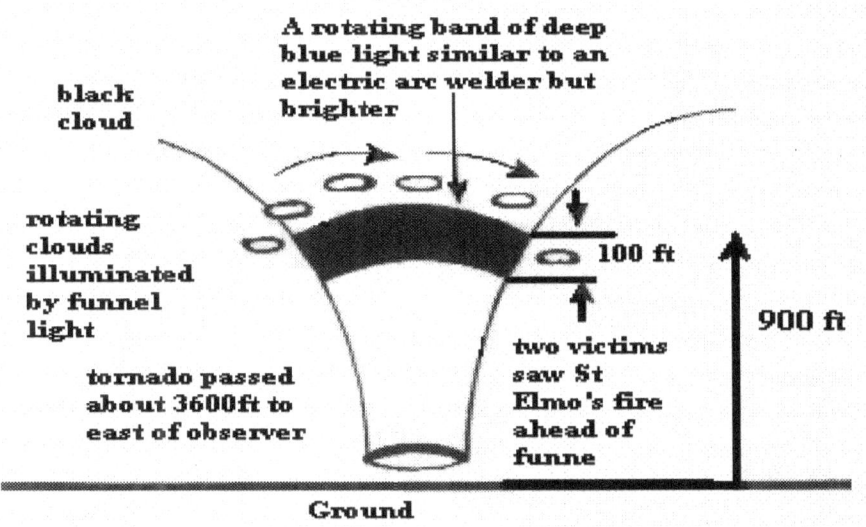

Figure 18 : The Blackwell tornado and the peculiar luminous torus

would be "shorting out". The answer lies in the vortex burner theory, which, in principle, easily surmounts this problem. Gas would be continually "piped" into the burning "furnace" of the vortex-breakdown zone amidst the water spray consisting of droplets which, though cooling certain zones would not quench a fiercely burning fireball. In this situation there should be plenty of steam generated as water is sucked up the spout impacting on the main burning region. How much quenching would take place is an open question needing further investigation. Then of course some of the water would be centrifuged out in a radial direction into cylindrical concentric sheaths like that cited in Kangieser (1954) away from the burning region.

The enigmatic "Andes light".

As I searched for ball lightning references I came across another extraordinary phenomenon, referred to, in the scientific literature, as the "Andes glow" or "Andes Light". The Andes light is a generic term for a series of unexplained light displays that take place on the tops of mountains, not only in the Andes, but also along other mountain ranges, such as in European Alps and Mexico. A common explanation of these lights is that they are caused by electrical displays, and like ball lightning, these lights have also been seen during fair-weather electric conditions which poses a real problem for electrical theories.

Markson and Nelson (1970) in *Weather* mentioned that low humidity was a common meteorological condition associated with the lights. They postulated that the lights may be due to a large scale St Elmo's fire effect. In effect, this would mean a high voltage corona discharge from the sharp crests and tops of mountain peaks. To test this idea out the authors carried out field work on the top of two Mexican volcanoes called Popocatapetl and Ixtacihuatl situated not far from Mexico City. They measured the electric field strength along the crest of the mountain tops and found it to be high. They suggested that corona discharge events could take place, but they did not have any evidence for this.

Although electrical effects may explain some of these lights, one weakness of the electrical interpretation, as Markson and Nelson pointed out, was that such corona discharge theories are seemingly unable to account for the beams, rays and haloes reported. Great "beams" have been seen miles out to sea. The description of such beams going all the way out to sea may be ball lightning in the form of a burning vortex acting as a "beacon with a search light". This is like a lighthouse beaming out from a mountain top. For large vortices the amount of light generated could be easily enough to produce such a beam of light.

The idea of a burning vortex fireball acting as giant torch or search light to cast an appreciable amount of light on to the sea surface became all the more credible to me, when I found the following ball lightning event. At the first international symposium on ball lightning a search light description was given by a Chinese scientist, Zou (1988), who reported on a ball lightning or UFO-like object in the skies over North West China. He described the object as being the shape of a basket ball and emitting light like a searchlight beam. The 1979 Kaikoura UFO lights similarly displayed this searchlight behavior.

Explanation of haloes

How are the halo effects on the Andes mountain range to be accounted for within the theory? The vortex fireball high up in sky can be approximated as a spherical light source which would radiate isotropically i.e. in all directions. Just as lunar ice crystal haloes are seen on cold winter's night from the refraction of light, fireball haloes could also be generated from radially-emitted light through airborne dust, water droplets or ice crystals. Coronas resemble haloes but they are caused by diffraction of light. These could also be observed around the fireball under foggy conditions. Finally combustion studies have shown that flame haloes can be produced (Gaydon and Wolfhard, 1953). There is clearly a need to find out what type of halo is involved in any specific fireball event. In some cases both types of halo could be present.

Explanation of searchlight beams

A glowing sphere is an isotropic source of light sending out light in all directions. How can the searchlight beam description, which implies a non-isotropic source, be accounted for? The explanation is basically the same reason why you can see sunbeams coming through a crack in a wall into a dark and dusty room. You see the beam because of the reflection off the dust particles or water droplets (sometimes like a cloud) which act like tiny mirrors which scatter the light. In the same way particle sheaths concentric with the spin axis of the vortex are formed from the centrifuging action in the vortex column illuminate and scatter light to the observer. Dust particle sheaths have been observed in nature in dust devils by Snow (1984) and others. Light is reflected off these particles all the way down the nocturnal vortex column creating the effect of a search light. It logically follows that the searchlight beams, seen in the Andes effect, and other cases of ball lightning, are really the illuminated funnel or "beam" running parallel to the main axis of rotation the vortex. If an unexplained light column is observed at night at an angle of 45 degrees to the sea surface, then one could infer that the vortex funnel is also inclined at that angle. Conversely when particles sheaths are not present there is then a lack of light scattering particles in the vortex. Therefore no beam is seen.

There is some evidence pointing to the idea that illuminated dust funnel "beams" can explain the observation of Vonnegut and Weyer (1966). They presented a photograph of some illuminated columns of light. The date was 11 April 1965, and the photo was taken at the time a series of tornadoes swept through Toledo. The columns of light were thought by the authors to be illuminated tornado funnels because the camera, which took the photographs of the illuminated pillars, actually pointed in a direction which was found later to have actually intersected the path of the tornadoes. A suggestion was made by the authors that this luminous phenomenon was directly associated with these tornadoes which were thought to be presters-possibly illuminated by electrical discharges. Hence this photographic evidence could be the first of its kind of Andes-like searchlight beams.

Further reports of searchlights and illuminated pillars of lights in connection with tornadoes were cited in the same report. One afternoon in Campaign, Illinois during the summer of 1942, W.S Houston saw a beam, like a searchlight which extended down from the clouds to the skyline over a time interval of one to one and half minutes. The beam was then suddenly replaced by a tornado funnel which came into sharp focus. D.B. Munroe reported that several witnesses saw a tornado column lighted on the inside. Similar such descriptions have been seen with UFO reports. Observers mention "rays" or beams which lengthened as they descended towards the ground. The explanation in these situations could well be illuminated vortex funnels.

Although such funnels were illuminated one must look high up to discover the source of the light. The nature of this light source should coincide generally with the vortex-breakdown zone where the combustion process would take place. The air speeds in the funnels (several tens of meters per second) which would easily exceed the combustion flame speed, (around $1ms^{-1}$ for standard atmospheric pressure). This would explain why the funnels would appear to ground-based observers not to be generating their own light.

Anomalous heating effects from tornadoes.

Unusual heating effects due to tornadoes have been reported in the scientific literature. Silberg (1966) wrote a paper that was published in the *Journal of Atmospheric Sciences*. In the introduction of his paper the author wrote that the burning and scorching effects of tornadoes have been recorded as early as 1839. The tornado funnel burns, dehydrates, and scorches various objects in its path.

Silberg went on to examine two hypotheses by Montgomery and Vonnegut before proposing his own ring current model. Montgomery's hypothesis was that a tornado could heat and burn objects, like trees, with a "hot wind". In an alternative theory, Vonnegut (1960) believed that corona discharges or St Elmo's fire took place at the base of the tornado and contributed to the burning.

Silberg found both explanations problematic. He stated that the hot air hypothesis did not have direct field evidence. For example, there have been no such reports of a sufficiently hot wind associated with a tornado to scorch vegetation such as trees. He believed that given the short contact time of a tornado the heat from such a hot air blast would not be enough to dehydrate tree stumps.

Another problem for these theories is what Silberg termed, a "thermal discontinuity". An example of thermal discontinuity is when tree trunks are burnt on the side facing the oncoming tornado while the other side is untouched. He wrote that this discontinuity is not consistent with any known heat conduction process and he believed that some other more sophisticated process must be proposed.

The corona discharge hypothesis fares no better than the hot air hypothesis. Corona discharges may be able to explain the smell like that of sulfur, but the heating effects, such as the dehydrated tree stumps, are very difficult to explain. The reason is that the corona discharge is a low heating power phenomenon and therefore it would have little if any significant heating effects associated with it. Wood requires something like 15-20kWm-^{2} from a radiant source. In addition the corona discharge requires an electrode of small curvature to establish the high electric fields needed for the electrical breakdown of air (about 30 MVm-^{1} for dry air)

The most obvious mechanism to account for the asymmetric scorching, like the burning of one side of a tree, would be via a radiative transfer process. The energy required to burn the tree is transported by infra-red electromagnetic radiation to the tree trunk. This is like holding a slice of bread up to fire so that only one side gets toasted. It was therefore no accident that Silberg postulated a radiative transfer process in the form of a ring current high up in the tornado to "toast" a tree. The extent of scorching and burning of the tree would depend on the tree species, the water content of the trunk, the bark and so on. I agree with this radiative approach, but I see no reason why instead of an electrical ring current as a source, a tornado furnace with the combustion process in vortex-breakdown could equally serve as a powerful heat radiation source.

Black circles and reduction of metallic ores

Another mode of burning, other than radiative transfer, is by direct contact. As the vortex-breakdown fireball zone moves down and makes contact with the ground. Combustible material like grass, broken branches and trees where the fireball touches this material it could ignite and burn it or simply melt it. A fireball impacting on grass would produce a characteristic circular area scorched on the grass and dehydrate the soil and clumping it together. In soil areas containing metallic ores and with both high flame temperatures and a reducing agent, such as carbon, could reduce metallic ores to the metal. Shiny metallic globules might then be found in areas where the fireball vortex touched down. For some lead ores the flame temperature would only need to be around 400 degrees Celsius for a reduction reaction to the metal. Other ores would need higher temperatures. Even non-metallic oxides could be reduced. Silica sand when heated to a hot enough temperature could also produce natural glass globules or fibers.

St Elmo's fire.

St Elmo's fire, also called "corona" is the light emitted from a low amperage electrical discharge into air. To produce this effect, high electrical fields need to develop at the surface of a sharply curved metallic conductor. This effect can be reproduced in the laboratory using high voltage applied to two electrodes. St Elmo's fire takes place naturally in mountains during a thunderstorm generating a corona glow on sharp objects, like ice axes. Corona remains in the vicinity of the high electric field of the object and attached to the conductor. This stringent requirement is at odds with the reported motion of ball lightning far from a high voltage conducting cable or a sharp object. Ball lightning is a more mobile and independent entity than St Elmo's fire.

It is quite likely that a number of genuine St Elmo's fire events have been misidentified as ball lightning perhaps because a corona discharge can assume a spherical glow.

There are ball lightning theories that use St Elmo's fire as a central concept. However St Elmo's fire is a low current low energy phenomenon and so it cannot account for high energy ball lightning events. Such a theory does not explain rotation or the splitting of a single ball lightning into two or more balls and recombine. Coronal discharges generate very little heat. This is natural since it is a low current discharge (eg milliamp range). On the other hand there have been substantial mechanical and heating effects with ball lightning, including plenty of reports of ball lightning exploding. In nature St Elmo's fire requires a high electric field in the atmosphere, which typically takes place during thunderstorms. But such a theory runs into another problem. There are a significant number of ball lightning sightings in fair weather with a much lower electric field conditions (i.e. $100Vm^{-1}$.)

St Elmo's fire changing into ball lightning.

I discovered another odd connection between ball lightning and St Elmo's fire. There are reports of St Elmo's fire transforming into ball lightning. How is this possible? Consider a vertical metal rod sharpened to a point. Imagine also a spherical coronal discharge around the point of the rod. Suppose that there exists a vortex-breakdown region vertically above the rod, generated by a draught of air containing a quantity of natural gas, as it flows over the rod. Vortex genesis could arise because the rod acts as an obstruction to the air flow. Vortex creation process at a metallic rod could be analogous to how a swirling eddy forms behind a rock. In electrically-active conditions the sharp point of the rod would be hissing with the sound of a corona discharge (i.e. St Elmo's fire). All three ingredients of the vortex burner theory are then present and the fireball bursts into life. A spark from the rod or a heat source then ignites the gas so combustion would proceed. To an observer a mysterious transformation would take place with St Elmo's fire changing into a dynamic fireball. Once this rare transition is made an independent vortex fireball could detach itself from the rod, perhaps from a fluctuation in the prevailing air current flowing past the rod.

The above transformation could arise in a number of contexts especially in association with metallic conductors, like the wings of aircraft where there exist trailing vortices created as the plane moves through the air. In that case the ball lightning could be seen to "follow" the aircraft. This could be one reason why ball lightning appears to give the impression that it pursues an aircraft.

Some writers on ball lightning claim that corona can move along any metallic conductor, such as a cable line. But this may only happen if the cable is at a voltage high enough for the electrical breakdown of air. An alternative scenario, though very much less likely, is that a high electric field develops around the wire as a result of high external electric fields from stormy weather.

A Hamiltonian ball lightning carrying a scarf.

Unlike St Elmo's fire, ball lightning has been seen moving on low voltage or ordinary wires or cables where it is able to suddenly detach itself and move aloft. One such event took place in Ireland in 1852. The original report was from Williams and Packenham (1872)

"One of Sir William Hamilton's sons, who was fond of experimenting, had a telegraph wire running the length of a shrubbery, as far as the lodge; whether from accident or in the natural course of things I don't know, but an accumulation of the electric fluid, I suppose, passed along the wire from the Observatory and down to the lodge, where it left the wire, passed along the ground a distance of some yards, then rising over a hedge, on which some linen was bleaching, picked up a cravat and mounted high in the air with it, carrying it a distance of about half a mile. I saw it myself like a ball of fire with something white hanging to it; but I could not judge very well of its height."

This ball lightning climbed high in the air and carried a scarf off the ground. Where is such a lifting force going to come from a self-contained plasma ball? Electrostatic effects would hardly be strong enough, and if they were, the scarf would be attracted to the hot fireball and quickly burnt. It would seem more practical to imagine the existence of a vortex with the lift originating from an upwardly-directed current of air.

There is no mention in the above account of the scarf burning. If

137

the scarf was carried by a vortex, a set of forces would have acted on the scarf whereby the downward gravity force on the scarf balanced the lift force of the vortex at some location below the burning zone thus preventing complete combustion of the scarf for the half mile. Scorching of the scarf could have taken place but it appears no one was apparently close enough to witness any such effect.

Bach's research on Gorgones from Volcanoes

One author who believes the original conception of St Elmo's fire is far more dynamic than the usual coronal discharge phenomenon is an American geologist and researcher, E. Bach. He conducted a number of expeditions into different countries to record moving fireballs hovering around during volcanic activity (Smirnov, 1994). Bach (1993) claimed the term "St Elmo's fire" was used by early fishermen and sailors to describe moving fireballs, rather than a corona discharge. In Appendix 2 of his Bach asserted that the original St Elmo's fire was seen in the vicinity of volcanoes especially underwater volcanoes. They were said to emerge and translate upwards from these volcanoes after having bubbled from the sea. They were able to move down on top of ships. These lights were either red or bluish-green and were seen in connection with the Italian volcanoes, such as Vesuvuis and Epomeo. They were also seen by Philippine fishermen.

Bach called these "St Elmo" volcanic fireballs, "gorgones". According to my Oxford dictionary gorgones means "terrifying women". This St Elmo fire may be explained with the help of the vortex burner hypothesis. It is significant that the St Elmo's fire was seen in connection with volcanoes since this is where I would expect vortex fireballs to be found. Such volcanic vortex fireballs could use natural gas as a combustible fuel gas seeping from fissures in and around these underground volcanoes. The observation of fireballs bubbling from the sea is certainly consistent with the escape of gas from a seawater-to-air interface associated with earthquake light production.

A new understanding of fireballs from volcanoes

UFOs and ball lightning from volcanoes is now understood in broad terms at least. As I have said earlier volcanoes provide a hot convective updraft. With a shearing winds vortices can form downwind. Any natural gas and especially hydrogen sulfide (auto-ignition temperature of 270 degrees Celsius) ignites in the hot surrounds of the volcano. So it is now not so strange that Bach thought that UFOs came from volcanoes. But my unified theory suggests that these are not the only locations on the globe-but they are certainly very active locations.

Black ball lightning.

"Black ball lightning" is a term used in William Corliss's book entitled, *Lightning, Auroras, Nocturnal Lights and Related Luminous Phenomena-A Catalog of Geophysical Anomalies*. I was curious to know why ball lightning, which is usually a luminous phenomenon, could be black. There must be some mechanism to keep black material in roughly spherical zone. How could plasma theories, which are reliant on hot plasma and electro-magnetic fields, explain black ball lightning? Corona discharge theories, microwave theories, even many combustion theories are also dependent on luminosity being present. Flame combustion theories fall down, even if smoke was produced, since there must be some mechanism to confine the smoke. Black ball lightning might then consist of remnant material, notably smoke confined in a quasi-spherical region.

The following case in Corliss's book shows that smoke can manage to get into a roughly spherical configuration. I strongly suspect a smoke-filled recirculating laminar vortex-breakdown as the correct explanation. On August 28, 1928, at Peoria, Illinois a lightning stroke hit a tree and then at about two meters from the tree, a ball of smoke appeared less than a meter above the ground. The ball was about 46 centimeters in diameter and was spherical. The smoke's color was yellowish-brown like smoke from burning straw. The smoke from the ball then diffused into the surrounding air. Smoke rising within the recirculation system of vortex-breakdown would act as a momentary tracer to delineate a spherical region.

Black ball lightning has been seen inside buildings. One such event took place on the 9th of December, 1897 at Castelgandolfo, Italy and was reported by Mathias (1926).

"After close lightning, a black globe 25-30 cm in diameter appeared in a kitchen. Seemingly formed of dense smoke, it scattered sparks and escaped through an open window."

The reference to sparks suggests the black globe was composed of combustible material like wood from a wood fire which has gone out but left smoke and a few embers. In the case of ball lightning the vortex-breakdown is equivalent to the hearth while the ash and some glowing embers are larger particulate material recirculating within vortex-breakdown.

I found the above black globe case similar to the report by Chaggar (1982). A burning disc-like object lost its luminosity and ended up becoming smoke which did not disperse but moved out into the kitchen. Unfortunately additional characteristics like the geometrical form of the smoke or whether it was moving were not mentioned in the report. It is worth stating that in forced convective flows, such as in the 0.7 meter vortex generator, smoke can enter the vortex from the along the boundary layer. Some portion of the smoke goes into the vortex-breakdown bubble and is swept up to exit the ducting never to return. But a hovering vortex in the atmosphere has an overall convective air flow pattern where air could leave the vortex return and feed back into the vortex. Hence the report of Chaggar might be accounted for, but more details of the case would be required to settle the matter. Smoke could travel in such a flow. From this point of view the current generation of vortex generators is unrealistic in modeling unconfined vortices seen "hovering" in the air.

Wallace and Teng's (1980) report may contain a possible reference to black ball lightning. In a Chinese province fireballs were seen at night and smoke balls during the day. Fireballs were observed where river beds met and along fracture zones. These river bed smoke balls may be related to ordinary combustion fireballs like a petrol fireball originating from rapid combustion or they may be vortex fireballs burning natural gas coming from fissures. Unfortunately there is not enough information in the above description to say one way or the other. But the fact that there was a high frequency of fireballs rather than amorphous flames may point to vortices.

Bead lightning ("pearl" or "rosary") lightning

Bead lightning is an unusual event that takes place after the main lightning discharge has decayed away. An observer sees a string of globular lights (the "beads") resembling the form of "pearl necklaces" or Catholic rosary beads left behind after a lightning stroke. Hence bead lightning is sometimes dubbed "pearl or rosary lightning". Lewis (1988), from Abu Dhabi, in the United Arab Emirates, reported in Weather, *on* the long-lived luminosity of a line of luminous balls well after a cloud-to-cloud lightning stroke had ceased.

Karl Nickel, a German ball lightning researcher, used the book *"Ball lightning and Bead Lightning"* of Barry (1980) as a source for his comments on bead lightning. The various names for bead lightning include: pearl lightning, perlschnurblitz, and clair en chaplet. Reports on bead lightning are rarer than ball lightning, but it is considered by some to be a type of lightning.

The problem with bead lightning, like ball lightning, is that there exists no widely accepted scientific explanation. At the beginning of my investigation in 1988, when I first read Barry's book, I thought that bead lightning might be a form of ball lightning rather than ordinary lightning, in the form of a "string" of ball lightning. I speculated that if a solution to ball lightning could be found then this may shed light on the bead lightning riddle as well.

Vonnegut (1960) briefly suggested that bead lightning and glowing fireballs, seen in connection with tornadoes, are either the same, or closely associated with ball lightning. So it cannot be ruled out that bead lightning is actually a series of ball lightning objects linked in a chain. Professor Nickel (1988) also suggested this possible explanation of bead lightning. In Nickel's opinion bead lightning could easily be viewed as a string of ball lightning spaced out along the previous path of a lightning stroke. He stated that such an idea has been suggested several times in the literature.

I have found indirect evidence that points to bead lightning as a string of vortex burners created by a lightning stroke. For instance, the diameter of the individual "beads" of bead lightning are within the range of the mean diameter of ball lightning. There is also evidence that the conditions in which bead lightning has been observed may be conditions suitable for the creation of vortices. Powell and Finklestein (1970) in their paper entitled, *"Ball Lightning"* published in *American Scientist,* described how a destroyer dropped depth charges in Chesapeake Bay in 1957 to produce water waterspouts. These waterspouts were filmed and analyzed. In one trial, lightning struck a water spout and created ten to fifteen balls of light. Each ball was estimated as 30 centimeters in diameter and lasted about 0.2 seconds after the flash of each stroke-there being three strokes in total. The "beads" then decayed away.

I examined the photograph showing the beads. It was hard to distinguish some of the individual beads, but I counted seventeen in the photograph. The important aspect was that the photograph, upon close examination, showed a few individual "beads" that were actually connected. A twisting lightning stoke could have produced a series of vortices all along the length of the lightning channel. Perhaps the water spout acted as a preferred conduction route for the lightning strokes. The threads connecting the beads are identified as combustion flames. Multiple vortex-breakdown of a single vortex is possible with combustion taking place at each bead site.

Another event which seems to show a close association of bead lightning to ball lightning is contained in Singer (1971). Boll in 1918 reported on a ball lightning which appeared after what looked liked bead lightning. This occurred after a series of cloud-to-earth lightning strokes were repeated along the same lightning channel. A glowing ball formed at the top of the channel just beneath a cloud and descended. Soon after, the process was repeated. Perhaps the process that gives rise to the beaded lightning could also have produced the glowing ball.

Also intriguing is an even rarer type of bead lightning, in the form of a set of isolated beads that are noticeably strung together in a straight necklace formation. Cade and Davis (1967) page 81, called this phenomenon, "chain-shot" lightning. Because there are clearly defined luminous connections between consecutive balls, a multiple vortex-breakdown phenomenon may be involved.

Whirlwinds of smoke and fire.

How can some whirlwinds be on fire and emit smoke in the combustion process? Surely the fire would be put out by the rushing air of the whirlwind. The existence of such fiery whirlwinds would seem unexplained. But vortex-breakdown with its stagnation zones now makes this a real possibility.

Corliss (1982), on pages 120 to 122, cited nine episodes of fiery whirlwinds. The observations of this type of phenomenon demonstrate that these whirlwinds have the characteristics of ball lightning observations. The observers clearly recognize the existence of: the whirlwind component, its accompanying mechanical effects, associated fire and smoke, and combustible material. It is this material that was actually seen to be drawn up into the vortex and burnt. There were loud noises, including crackling and explosions. The vortex burner hypothesis accounts for these observations. Corliss (1982) reported a case in *Symond's Monthly Meteorological Magazine*, edition number 4. On page 123 a type of prester was reported in Ashland, Tennessee in 1869. This phenomenon exhibited a number of characteristics of what I would expect of a vortex burner. One of these features is the capacity of the vortex to lift material off the ground into the burning vortex. The following account, movie-like in its description, illustrates the remarkable advance of a whirlwind pillar of fire in fine weather. The effects of burning on vegetation match the anomalous heating effects of tornadoes that I have previously discussed.

" ... *a remarkably hot day ....a sort of whirlwind came along over the neighbouring woods, taking up small branches and leaves of trees and burning them in a sort of flaming cylinder that traveled at the rate of about five miles per hour, developing size as it traveled. It passed directly over the spot where a team of horses were feeding and singed their manes and tails up to their roots; it then swept towards the house, taking a stack of hay in its course. It seemed to increase in heat as it went, and by the time it reached the house it immediately fired the shingles from end to end of the building, so that in ten minutes the whole dwelling was wrapped in flames. The tall column of traveling calorific (i.e. energy/fire) then continued its course over a wheat field that had been recently cradled, setting fire to all the stacks that happen to be in its course. Passing from the field, its path lay over a stretch of woods which reached the river. The green leaves on the trees were crisped to a cinder for a breadth of 20 yards, in a straight line to the Cumberland. When the "pillar of fire" reached the water, it suddenly changed its route down the river, raising a column of steam which went up to the clouds for about half-a-mile, when it finally died out. Not less than 200 people witnessed this strangest of phenomena.*

A number of obvious inferences may be made about this phenomenon from the standpoint of the vortex fireball theory. The burnt leaves suggest that the burning cylinder had a diameter of around 20 yards (i.e. about 20 meters). The whirlwind in moving over a body of water was able to not only suck water into the vortex but to also continue burning. This burning of a combustible fuel turned the raised water to steam. What the fuel for the fire was is not clear. Though a combustible substance could have been drawn into the vortex, along with the water, it does not necessarily have to be natural gas. So long as there is some burning material inside vortex-breakdown, the sustained luminosity and heating of the water can be accounted for. Combustible material like burning wood can be drawn into the vortex and suspended above the ground from a balance between the gravity and the lift force associated with the vortex.

Corliss included another amazing observation within his "whirlwinds of fire" category, namely the Newbottle whirlwind of November 30th, 1872 that took place at Banbury, England. This object was reported by Beesley (1873) and described as a "huge revolving ball of fire" in the following report:

"About 12 o'clock we had a heavy storm of rain and hail, in the middle of which there was a very vivid flash of lightning, with almost instantaneous thunder of a very peculiar rattling sound. About five minutes afterwards, as I was leaving the house, my gardener called me to come and see the ball of fire. I was unfortunately half a minute too late, but I have seen four persons who saw it from different points, and who all agree they heard a whizzing, roaring sound like a passing train, which attracted their attention, and then saw a huge revolving ball of fire traveling from six to ten feet off the ground. The smoke was whizzing around and rising high in the air, and a blast of wind accompanied it, carrying a cloud of branches along and destroying everything in its way.... Where it first began the breadth of ground traveled over was very narrow, but increased as it proceeded, till in the last field the debris covered a space quite 150 yards wide, and here it seems to have exhausted itself, as all witnesses agree that the ball of fire seemed to vanish at this spot without any explosion. Here the ground had been cut in places as if by a cannonball, but I could find no cause for this, and I saw no signs of fire on its route...William Marshall, gardener at Newbottle Manor, was returning from stables to the house. He heard a noise like a long railway train crossing a bridge, and saw leaves and branches whirled into the air above the Spinney, and immediately afterwards 'a dark ball, as big as a carriage', and sending up a 'a cloud of smoke' come out of the trees with a shower of branches, and roll 'over and over', down the hill in the direction of the bridle road; the cloud of smoke at the same time whirling 'round and round' with a 'buzzing noise'. He

distinctly saw sparks of a red color emitted from the ball about six feet from the ground, and this is confirmed by another man, William Jilson, of Astrop, who, from a field on the west, saw fire and ran away affrighted."

The scientific information in this vivid description suggests a vigorous whirlwind with a revolving fireball situated upon the axis of rotation of the vortex while at the same time burning organic debris. The huge revolving fireball is essentially a burning whirlwind in a state of axisymmetric vortex-breakdown. The lifting of leaves off the ground, the smoke carried up, the buzzing noise, and the cutting into the ground all point unequivocally to a natural vortex.

In the Monthly Weather Review of July 1881 (Issue 9, No.6, and page 9) there was yet another remarkable report of a twister on fire. In this case the small twister was seen to form in a corn field. There was a fire in the center and the identification of a kind of "sulfurous vapor" which no doubt formed the fuel component feeding into the fiery vortex in the report below:

"...a small whirlwind, about 5 feet in diameter and sometimes 100 feet high, formed over a corn-field where it tore up the stalks by the roots and carried them with sand and other loose materials high into the air. The body of the whirling mass was of vaporous formation and perfectly black, the center apparently illuminated by fire and emitting a strange 'sulphurous fire' that could be distinguished a distance of about 300 yards, burning and sickening all who approached close to breathe it . Occasionally the cloud would divide into three minor ones, when the whole mass would shoot upwards into the heavens.

The mention of a sulfurous vapor is contained in several references to tectonic activity. Thomas Gold in his book, *Power from the Earth* has a collection of reports that almost certainly suggests that prior to some earthquakes there is a substantial emission of gas which creates a kind of fog that is often smelt as a sulfurous vapor (it could be that in this case hydrogen sulfide is burning in the vortex to produce sulfur dioxide as a byproduct).

... The next day but one before the first earthquake was darkened from morning to night by thick fog and divers persons perceived a sulphurous scent.

The basis of my theory is dramatically shown in these fiery whirlwind accounts. These observations provide the best evidence yet. There seems little doubt concerning the existence of the phenomenon on which this book is based. A revolving fireball situated upon the axis of rotation of the vortex has to be explained. The favored explanation is that the fireball is identified as combustion in vortex breakdown. The secondary vortex effects are the whirling leaves, the smoke carried up, the buzzing noise, and the cutting into the ground.

Anomalous "meteors"

Several cases of ball lightning have historically been misreported as actual meteors. Conversely several "meteors" have been erroneously interpreted as ball lightning. The hypothesis that stones fall from the sky was originally greeted with great derision. Antoine Lavoisier, of the French Academy, and a leading scientist and chemist, pronounced that stones could not have fallen from the sky simply because there were no stones in the sky! He was referring to the 1772 event where the people in a French town called Luce saw rocks that came from the sky. In 1794, the physicist Chladni concluded, with supporting evidence, that meteors were extra-terrestrial but his claim was largely ignored. Then on April 26, 1803 around 2,000 meteorites fell at L'Aigle in France. The event was recorded in a paper by French physicist Jean Baptiste Biot (1774-1862). He convinced fellow scientists that there really were rocks falling from the sky.

Once it had been established that meteors stones really do fall from the sky this led to a number of rival meteor theories. One such terrestrial theory of Benjamin Franklin sought to connect lightning with meteor production. Apparently this was a time when it was fashionable to form electrical hypotheses for many natural phenomena. But this "thunder stone" theory was finally killed when it was discovered that meteors consistently came from preferred locations in outer space. The Leonid meteor storm of November 12-13, 1833 marks the birth of meteor astronomy because Professor Olmsted of Yale College had evidence that meteors appeared to originate from the constellation of Leo. The meteor hypothesis and its explanation finally gave birth to the science of meteors.

I previously mentioned that some ball lightning events have been misidentified as meteors. The scientific literature contains several references to these anomalous "meteors" that may have been incorrectly identified in the past. The illustrious Dr. Clyde Tombaugh, discoverer of Pluto also clearly differentiated standard fireballs from an unknown fireball(UFO).

"I have seen three objects during the past seven years which cannot be explained away as Venus, nor optical phenomena of the atmosphere, nor meteors, nor aircraft. I am an expert on astronomical observation and I have sighted eight green fireballs

moving in a quite different way from the usual fireballs."

The trajectory of these fireballs is quite different to the usual meteor path. They can have very low-angled paths, with sometimes sharp changes in direction. Ordinary meteors typically possess a parabolic motion. On the other hand these strange meteors have the ability to uncharacteristically rise up in the atmosphere and travel in formation. Schofield, cited in Corliss (1982), reported that in 1904 of a special meteor observation which was witnessed by three people. The event took place in the North Atlantic Ocean on February 28th, 1904. The report below showed that the "meteors" may have been a set of vortices which moved together as a result of the vortex splitting phenomenon.

"The meteors appeared near the horizon and below the clouds, traveling in a group from northwest by north (true) directly toward the ship. At first their angular motion was rapid and color a rather bright red. As they approached the ship they appeared to soar, passing above the clouds at an elevation of about 45 degrees. After rising above the clouds their angular motion become less and less until it ceased, where they appeared to be moving directly away from the earth at an elevation of about 75 degrees and in a direction west-north west(true). It was noted that the color became less pronounced as the meteors gained in angular elevation. When sighted, the largest meteor was in the lead, followed by the second in size at a distance of less than twice the diameter of the larger, and then by the third in size at a similar distance from the second in size. They appeared to be traveling in echelon, and so continued as long as in sight. The largest meteor had an apparent area of six suns. It was egg-shaped, the sharper end forward. This end was jagged in outline. The after end was regular and full in outline...The second and third meteors were round and showed no imperfections in shape....When the meteors rose there was no apparent change of relative position, nor was there at any time evidence of rotation or tumbling of the larger meteor. I estimated the clouds to be not over one mile high. The near approach of these meteors to the surface and the subsequent flight away from the surface appear to be most remarkable, especially so as their actual size could not have been great. That they did come below the clouds and soar instead of continuing their southeasterly course is also equally certain, as the angular motion ceased and the color faded as they rose. The clouds in passing between the meteors and the ship completely obscured the former. Blue sky could be seen in the intervals between the clouds. The meteors were in sight over two minutes and were carefully observed by three people, whose accounts agree as to the details.

Cade and Davis (1967) commented on these odd "meteors" and

described their form and behavior.

"There is a class of unusual meteors which are believed to be caused by fragments of comets (cometoids), in other words, by lumps of frozen water and gases. The theory applies especially to objects of substantial size, from a few inches to many feet in diameter, which do not plunge headlong into the earth as do meteorites (meteors), but slow down to very moderate speeds at altitudes of a few miles, and which exhibit certain peculiarities, such as fuzziness of appearance at low altitudes, wobbling in flight, brilliant light emission (particularly of an apple green color) and a total lack of solid remains....They have been observed not only to wobble but to change their flight path, and their trajectories when below a thousand feet are nearly horizontal."

Cade and Davis rightly drew attention to a list of inconsistencies between these special meteors and the usual meteor. The "total lack of solid remains" from chunks of the meteor that should have hit the earth and the very low altitude paths were noted. These characteristics would also apply to the famous Tunguska "meteor" crater that I will discuss in more detail. These characteristics along with other peculiar motion and changes in flight path suggest that a vortex burner fireball was involved.

Ol'khovatov (1997) described another case of an anomalous "meteor" observed in the town of Belinskii (formerly Chembar) in Russia on January 4th, 1886. I found the following clues in his report indicating a vortex fireball. A so-called "meteor" took a low altitude path over the town with an unusual gust of moving air associated with it. The object exploded and killed a horse. An earthquake was experienced 10 to 15 minutes later. The moving air was said to be directly connected with the meteor because the presence of wind was noted. The moving air and low speed are quite uncharacteristic properties for an ordinary meteor. The presence of wind or moving air is, of course, a characteristic feature of a vortex and definitely suggests the passage of a tornado through the town.

The Chembar meteor was seen before the earthquake. That a fireball was observed prior to the earthquake may be a mere coincidence or it may be consistent with the accounts of gas emitted before earthquakes, as mentioned in Gold's book. If gas was released then it could have been used as a fuel for the fireball and a low-angled trajectory is possible with a vortex fireball. Although these observations provide only circumstantial evidence the vortex burner explanation is promising and further research (if available) of the event would gain a better insight into this particular meteor event

A link on the website by a Russian investigator, Ol'khovatov about a peculiar 'meteor' directed me to an article by Docobo *et al* (1998) in an issue of *Meteoritics and Planetary Science Journal*. In 2003 I wrote a reply to the same journal indicating that this type of non-meteor fireball event might be accounted for within the vortex-fireball theory. My aim was to approach the meteor scientists to point out that some so-called 'meteors' have been incorrectly classified as actual meteors. I thought that at a fundamental scientific level, for the field of meteoritics, it would be obviously important to be able to correctly distinguish between these normal meteors and vortex fireballs. I thought that publication of a short article might alert meteor scientists to this differentiation.

A previous editor of the above meteoritics journal indicated that the reply could be published but it became a question of what form this would take. After inquiries I waited for the acting editor to reply back but I heard nothing. Perhaps they thought the idea was too controversial. It is poignant that the birth of "meteoritics" was not without its own controversy. The very idea that stones fell from the sky was considered preposterous at the time. Here is the unpublished article where Tunguska-like observational data i.e. the absence of any meteor remains yet destructive effects at ground level and an unusual trajectory are features of the supposed "meteor". I have since found out that my suggestion has been seriously taken up with. Dr Vladimir Svetsov (private communication) of the Russian Academy said that he had cited my theory to explain the Compostela event. In his own words, "...*I quoted your book recently in a chapter written for a report for Sandia Laboratories in connection with the Santiago de Compostella event that happened on 18 January 1994 (Meteoritics and Planet. Sc., 1998, 33, 57-64. Spalding wrote in this paper that the Spanish slow fireball could be caused by combustion of natural gas. I indicated that your theory provided a mechanism to produce the fireball...*"

Here is my short communication.

A reply to the paper by Docobo et al of the Jan 18 1994 Fireball over Santiago de Compostella, Spain

"*The paper by [1]Docobo et al gave a detailed account of what seems to be a type of non-meteoritic fireball showing some features similar to the 1908 Tunguska meteor but on a much smaller scale. This phenomenon was witnessed by no less than 34 eye witnesses. Several features were found by the authors to be inconsistent with the usual bolide description. These include the: very low almost horizontal trajectory, low speed, relatively long time of observation (1 minute), observed in a small area of the sky, large angular size, absence of meteoritic shock, and no meteorite material found at the site. However in contrast to the Tunguska event there is no observed rapid shift in direction.*

An onsite examination of a crater within 1 kilometer of the projected path of the fireball, measuring 29 by 13 meters and 1.5 meters deep showed no evidence of meteorite material and a strange location of soil and trees which suggests some kind one-sided ejection mechanism. Uprooted trees and soil were found only down slope. The authors concluded that these anomalies make up a profile which departs from the usual bolide description. There could be difficulties with the explosion-rising plume hypothesis. One problem is that a rising plume of gases with a strong vertical component and burning thereof has to somehow explain the very flat trajectory of the fireball and the continued stability of a roughly spherical fireball envelope with a small tail.

An alternative theory which retains the earth gases emission aspect, mentioned by the authors, was used to identify a set of puzzling fiery vortices[2] and the Tunguska event[3]. What witnesses of the 1994 Santiago de Compostella event may have observed is a combusting atmospheric whirlwind fueled by a natural gas plume from the ground possibly connected with earthquake activity, as the authors surmise. This hypothesis has the ability to account for sudden changes in trajectory and tubes or tails that extend from the bottom of the luminous region and a vortex can have a large mechanical energy.

The possibility that such burning atmospheric vortices might exist was hypothesized to explain unusual luminous tornadoes that held fireballs embedded on the main axis of the vortex. The key feature which enables a vortex to burn is the low air speeds in the reverse flow of vortex-breakdown which are required to be equal to or less than the fuel gas flame speed (in the order of $1ms^{-1}$). Low speed turbulent eddies in the core may be able to support some combustion zones away from the main vortex-breakdown combustion region and in main funnel. Furthermore an experiment confirming that combustion of gases like natural gas could take place inside the vortex-breakdown region of an atmospheric vortex has been demonstrated in the laboratory[4] using a purpose-built 0.9 meter diameter vortex generator.

As the 1994 'meteor' descended it was slightly inclined to the vertical in agreement with some observed vortices in nature. The tail of the vortex would be identified with combustion in the primary vortex core. The strong reverse flow of the vortex-breakdown region may have dug out the crater and flung the trees and soil to clear the path and with the vortex inclined at an acute angle in a down slope direction could explain the asymmetry of material flung from the crater area as illustrated in Figure 3 of the paper. An explosion connected with the vortex could have occurred if a combustible gas such as natural gas was present though the asymmetrical ejection of trees seems to rule this out. Such vortices are known to create trenches and so-called suction spots. The vortex fireball hypothesis predicts the complete lack of meteoritic material present at impact and low trajectories and may explain other anomalous meteor events which have been observed in the past."

Peter F Coleman

References
1. Docobo, J. A.; Spalding, R. E.; Ceplecha, Z.; Diaz-Fierros, F.; Tamazian, V.; Onda, Y. 1998, Meteoritics and Planetary Science 33,57-64,
2. Coleman, P.F., 1993, Weather, 48, No.1, 30.
3. Coleman, P.F., Identification of the Tunguska 'Meteor', 1997 Proceedings of the International Symposium on Tunguska, Russian Academy of Sciences, June 30 – 2 July.
4. Coleman, P.F., Abrahamson, J., 1999. Combustion Flame in a Tornado-like Vortex in a State of Vortex-breakdown, Eos,Transactions, American Geological Union, Spring Meeting.

Brontides

"Brontides" is a generic term used by Gold (1987) to describe unexplained and unusual explosive sounds in the atmosphere, usually in relation to earthquake activity. I believe brontides are most likely natural gas explosions. Ol'khovatov (1997) reported that whirlwinds and other wind action coincide with such explosions. Some of these brontides are gaseous explosions originating from a gaseous cloud or an exploding vortex fireball in its dying stages.

One possible scenario for a brontide vortex fireball is as follows. A vortex has the ability to accumulate natural gas from a wide area and ignite it to form a fireball. A decrease in the fuel-to-air ratio gas could put the gas mixture into the explosive regime and the fireball could end its life with an explosion i.e. the "brontide" explosive noise.

That vortex fireballs may be involved in some cases of brontides does have some support. Oh'khovatov (1997) reported on seismic activity in the vicinity of Arthur's Table on the Bala fault in Wales on January 23 1974. An explosion or brontide coincided with seismic activity and strange fireballs were seen in the area, both before and after the event. The direct reference to the occurrence of fireballs could turn out to be fiery vortices. The tectonic activity is certainly a pointer towards possible release of natural gas along the fault which would provide a vortex fireball with its necessary fuel.

A field investigation along the Bala fault could provide vital information to establish the occurrence of natural gas emission. The mechanism of vortex genesis in this area should also be examined and it may be related to air flows over an obstacle such as a mountain ridge coupled with an electrostatic discharge (even corona discharges) from a sharp feature in the craggy terrain.

The Tunguska Enigma

On June 30, 1908, near Lake Baikal in the swampy coniferous forests of the Tunguska River Basin of Siberia, a spectacular event took place. This was the mysterious and famous "Tunguska meteor" phenomenon recorded in numerous popular books and academic publications. Several theories have been advanced to explain this high energy "meteor" but there are difficult problems. Scientists are still perplexed about what really happened. A few have speculated it might have been a naturally-occurring nuclear explosion, judging from the destruction wrought. Antimatter from outer space was even mooted as a possible explanation of the Tunguska event.

Some years after the event, one noted Russian scientist decided to investigate. A meteor expert, called Dr Leonid Kulik, went to Tunguska in 1927 to investigate the incident. He was surprised because he expected a major impact crater but no such crater was found: not even meteoritic fragments. To add to the mystery, eighty million trees were uprooted and villages were burnt. A shallow depression was found but this was not at all typical of a meteor crater. The depression had a diameter of two miles with 200 shallow craters with steep sides. However, it is now accepted that these depressions are natural. These hollows are terrestrial swamps, karst holes and lakes (Vasilyev, 1996).

There is indisputable evidence that an explosion took place since a sound was heard 600 miles away and seismographs detected the event. Here was a real enigma. The meteor/comet hypothesis did not easily fit in with observations apparently connected with the event. Eyewitnesses at the time reported seeing a fiery ball with a tail, and also a glowing cylinder. This was noted in Berlitz (1987) but unfortunately the actual observation in his book did not have a source reference. Ol'khovatov (1997) described how the Tunguska object had a fiery tail. In fact, Ol'khovatov usefully assembled a good selection of vivid eyewitness observations that are consistent with the fiery vortex and vortex hypothesis rather than the usual meteor hypothesis. In one eyewitness account from Nizshne-Ilimskoye, the fireball, as it headed towards the ground, flattened into a "flying saucer" shape at ground level. When it "hit" the ground it changed into two fiery columns. A.K Kokorin, from the Kezshma Meteorological Station, recorded in the official register the observation of two enormous circles of fire which lasted for 4 minutes. A noise like a wind was heard. The descriptions of "fiery columns" and the "circles of fire" are consistent with the vortex burner theory as described in this book.

The motion of the object was very unlike a meteor path. Some eyewitness accounts report that the object came from the southwest and then changed course to the west. This description aptly fits in with the special "meteors" phenomenon I discussed earlier. Cade and Davis (1967) also pointed out this highly unusual motion. Some scientists have proposed a comet theory where a large ball of ice exploded above ground causing the shallow crater. But the eyewitness accounts do not tie in with this theory. Since there are problems with both the meteor and comet theories I would like to suggest an alternative hypothesis to explore. The Tunguska destructive effects are accounted for by a large vortex possibly a very powerful tornado that swept up natural gas from the swampy forest into the vortex-breakdown region and then ignited by lightning. This would certainly account for the above observations of the fireball, the tail, and the glowing cylinder and the change in direction which is most uncharacteristic of meteors and for that matter comets.

The Tunguska meteor made a tremendous noise. What may have happened is that a large tornado with an accumulation of natural gas exploded and the effects of this explosion would have been destructive on a large scale. The vortex explanation of Tunguska would mean that there would be no tell-tale solid particles left behind. The vortex would have dispersed along with the natural gas and its waste products. The effects of such a gas explosion would be all that was left. This would explain the fallen trees lying in a radial pattern, heating effects and the lack of material remains, radio-active or otherwise. There is additional evidence in favor of the vortex theory. Ol'khovatov (1997) indicated there were tectonic disturbances at the time of the Tunguska event and so natural gas may have been released into the atmosphere to fuel the fireball. Gold (1987) in his book reported cases where gas emission had taken place prior to an earthquake tremor.

Could it be that a quantity of natural gas flowed from cracks along the faulting zone in the Tunguska basin region to fuel this fireball? Is it any coincidence that the Tunguska epicenter took place at the intersection of fault lines (one connecting to Lake Baikal) and the Kulikovski volcanic crater (Kundt, 2001)? Kundt also pointed out that chemical irregularities reported by parties exploring the site were consistent with outgassing in tectonic zones. Further exploration of gas seepage in the basin could be carried out, if it has not already been done before by geologists, to test out this idea. If a large body of evidence of natural gas emission was discovered this would add even stronger support to the vortex fireball thesis as being the correct answer to what the Tunguska "meteor" was.

Remarkably, Ol'khovatov (1997) reported on a similar event to the Tunguska "meteor" that took place in Russia, near Sosovo on April 11, 1991. The Sosovo event may help in our understanding of what happened at Tunguska. In this case there were actual observations of fireballs. Fifteen to twenty years before the event there were tell-tale signs of tectonic activity such as ground deformation, the creep of piers. A strange explosion coincided with seismic activity. Fireballs and a bright fiery column of light were seen, as well as crater formation. There was also the distinct possibility that natural gas clouds accumulated prior to the sighting because some observers saw "foggy" clouds sink in the area where the explosion took place. Whirlwinds were also seen hovering over the site before the explosion took place." All this evidence is explained by the vortex burner theory.

The event can be explored further by use of the vortex fireball hypothesis. There is no doubt that an explosion took place. However, there may have been smaller explosions. So long as the tornado was able to survive and there was natural gas to feed this vortex then it could cut a path of destruction across the forested Tunguska region scorching uprooting trees as it went Remarkably in some eyewitness reports it appears that the fireball continued its path even after major damage to the area. (Vasileyev, 1996). This observation suggests that the source of natural gas continued to feed the fireball.

The reported inconsistencies between the meteor school explanation and evidence from the original eyewitnesses to the event and scientific field trips to the area were discussed by the Academician (higher rank than a professor) Vasilyev (1996) from Tomsk University. He discussed some of this evidence. Despite numerous expeditions to the Tunguska area no significant meteor explosion fragments have been found. Surveys reveal evidence of finely dispersed cosmic material could not be reliably distinguished from the usual background meteoritic activity.

As I have said there is also evidence that the Tunguska object managed to continue it pre-explosion trajectory which would seriously challenge some meteor or cometary hypotheses which begun with F. Whipple in 1934 and the Russian researcher, V.G. Fesekov. This is because most of these meteor and comet theories postulate either the complete disintegration by direct impact, or by exploding above the ground. In such a situation the cosmic object could not continue on its path.

The revival of comet theories commenced in the 1960s and seemingly explained the lack of any explosion crater and the absence of any significant parts of an exploded meteor. According to the comet theory having the comet explode kilometers above the ground does not explain the "vector" pattern of the fallen trees. Something more than a single epicenter at or near ground level (Vasileyev, 1996) is required. Furthermore some calculations show that if the Tunguska object was a comet, like Halley's Comet, the low cohesive strength of such a dirty snowball would mean that the object would have disintegrated much further up than is usually proposed.

When I first read about the Tunguska meteor there was insufficient data to propose the vortex fireball theory. But after reading the work of both Vasilyev and Ol'khovatov I became convinced that the meteor or comet school of theories was indeed problematic and the paradox would be resolved using the unified theory described here.

June 30th 1998 marked an international 90 year Jubilee Conference on the Tunguska enigma. Krasnoyarsk, the Siberian host city, is close to the Tuguska "epicenter". I sent off my proposed hypothesis to Dr Guiseppe Longo of Bolgna University on the 7th April, 1998. He is a researcher on the Tunguska phenomenon and hosts a web site on Tunguska. I sent an abstract to the Tunguska Organizing Committee in Krasnoyarsk (Coleman, 1998) on the 15 April, 1998. This abstract was published and is available at,

http://omzg.comcen-1.nsk.su/tunguska/en/newse/abstracts/kolem.htm.

IDENTIFICATION OF THE TUNGUSKA "METEOR"

P.F. Coleman
University of Canterbury, Christchurch, New Zealand

The June 30th 1908 Tunguska "Meteor" is identified not as a bolide fireball, but as a vortex fireball. The vortex fireball has been proposed as a basis for a theory of ball lightning. This alternative explanation seems to account for irregular fireball trajectory, tectonic activity and continued existence of the object after the explosive wave.

References:

Coleman P.F. An explanation for ball lightning? – Weather, V.48, No.1, 1993, p.31

Coleman P.F. Vortex-breakdown burner hypothesis of ball lightning. – In the book "Proceedings of the 5th International Symposium on Ball Lightning. 26-29 August 1997, Tsugawa Town, Niigata, Japan – 1997, pp. 176-182.

In this abstract I proposed that the Tunguska "meteor" be identified as a vortex fireball. The only alternative hypothesis that comes close is the "gas pouch" hypothesis proposed by a Rumanian professor. He invoked a natural gas explosion as the cause of the event, but a gas explosion on its own would not be insufficient to fully account for all the features of the Tunguska event that came to light.

PART II

UFOs

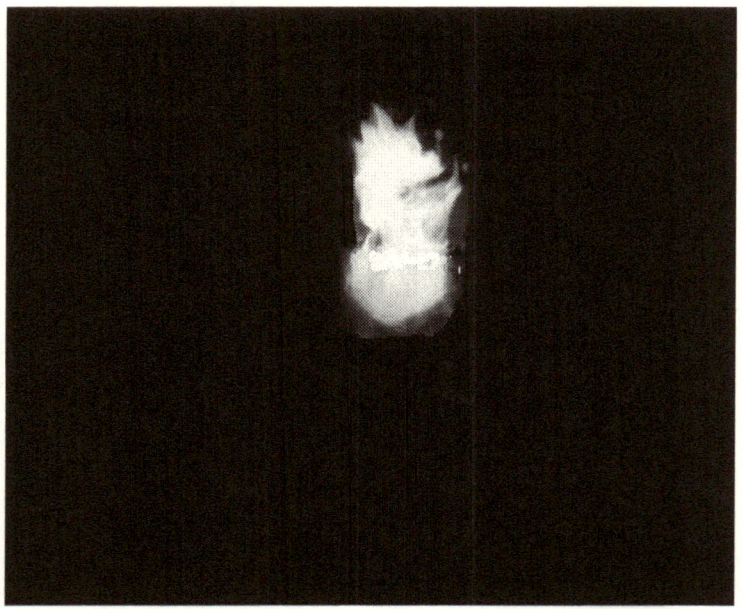

7
UFO theories

The formal "UFO" designation has become universally accepted the world over. The history of these mystery lights divides naturally into the pre-UFO and the current UFO eras. There is nothing special about the term; it stands for "Unidentified flying object" which could be a hubcap thrown into the air. However it is instructive to see when and where it first originated. Authors of the Mutual UFO Network Website (MUFON) believe the origin of the acronym "UFO" seems to have come from the Official investigation of the US Federal Government-Project Grudge (1949) and later, Project Bluebook. Captain Ruppelt, from the US Air force, himself, stated that he substituted "flying saucers" (from the 1947 Arnold sightings), with "UFOs" sometime between 1951 and 1953. The name change in these reports was perhaps, a piece of subtle verbal engineering designed to defuse the widespread notion of flying saucers. Instead these reports have done nothing to alleviate the idea that there has been a cover up, and that UFOs could be real and associated with alien spacecraft.

There are numerous cases of pre-UFO events, like the foo fighters and many more. The number of events reported in the UFO era would easily be greater, given the increase in population and technology to record such events. Possibly one of the earliest recorded sightings of fireballs was recorded in the Bible. The New Oxford Annotated Bible with Apocrypha (RSV) from the book of Ezekial (p 1000) reads,

"As I looked, behold, a stormy wind came out of the north, and a great cloud, with brightness round about it, and fire flashing forth continually, and in the midst of the fire, as it were gleaming bronze. And from the midst of it came the likeness of four living creatures. And this was their appearance: they had the form of men, but each had four faces, and each of them had four wings. Their legs were straight, and the soles of their feet were like the sole of a calf's foot; and they sparkled like burnished bronze. Under their wings on their four sides they had human hands...In the midst of the living creatures was something that looked like burning coals of fire, like torches moving to and from among the living creatures; and the fire was bright, and out of the fire went forth lightning..."

The *"burning coals of fire like torches moving to and fro"* in Ezekial's vision could well have been an actual eyewitness description of the motion

of ball lightning in a storm. Certainly the meteorological storm explanation of Ezekial's vision has been suggested before, in *Weather,* a publication of the Royal Meteorological Society of the United Kingdom.
Juliuis Obsequens in 90 BC saw a UFO hover over the town of Spoletum, Umbria. He recorded as;

"A globe of fire appeared in the sky at sunrise with a terrific noise, and burning. The globe was golden in color (and) fell to earth from the sky and was seen to gyrate...it rose from the earth, was borne east and obscured the sun with its magnitude."

Menzel (1953), a scientist and early twentieth century UFO debunker, cited Roger of Wendover who chronicled the history of England from 447 to 1235 AD. Roger described peculiar *"globes of fire"* which he personified as spirits may have been an early record of a ball lightning event. He wrote:

"As these globes of fire continued, now to rise on high and then to sink to the bottom of the abyss I observed that the wreaths of flame, as they ascended were full of luminous spirits.... glaring on me with their eyes of flame, and distressing me much with their stinking fire."

Roger's "globes of fire" are very like descriptions of modern day ball lightning or UFO sightings.
The author and film producer, Emmeneger (1974) used the UFO collection of the astronomer, Jacques Valle. On page 7 of his book, *UFO's Past and Present* he gave a brief account of strange luminous objects reported in Japanese history. On the 3rd of January in 1569 a flaming star appeared in the night sky. It was regarded at that time as an omen of serious political change predicting the fall of the Chu Dynasty. This phenomenon continued during the seventeenth and eighteenth centuries. In May 1606, fireballs were frequently seen over Kyoto. One night a revolving ball of fire was observed by many of the Samurai warriors. One UFO resembled a red wheel as it hovered near the Nijo Castle.
According to Emenegger, Contemporary Navy journals have reported numerous descriptions of fireballs at sea. On November 12, 1887, a huge fiery sphere was seen out at sea from the sailing ship, called the *Siberian.* The fireball object rose from the ocean and climbed to a height of sixteen meters. It has the ability to fly in the opposite direction to the prevailing wind and suddenly changed direction and flew towards the southeast. People saw this apparition for around five minutes.
There are so many names for the phenomenon; fox fire, will o' wisp and many others. The will o' the wisp type of light conjures up a image of

some sort of passive flame above a swamp, but there is more to it than this. Most commentators might not suspect a link between these types of lights and ball lightning and even some UFOs, but essentially they are the same phenomenon and governed by the unified theory I will describe in more detail herein.

Mystery fireball events have an ancient history. Just as thunder and lightning have been associated with gods and legends, fireball events have their own mythology. There are many interpretations of these lights prior to the rise of the monolithic project called science. In Japan these lights represented the souls (hitodama) departing the body. The Irish have various Gaelic names for these lights like *"teine biorach"* which means a fire or flame moving like a bird through the air (Clarke, 1988). The "fetch candle" is a name used in North of England folklore for a light, like a flame of a candle, which moved through the air at night. This light was believed to have accompanied a ghostly funeral. In South Hampshire the fetch candle was said to go out when the soul of the dying departs. A tribe of American Indians, the Penobscot, called this light, "Eskudait". It was said to be a fire creature and an omen of death. The Thonga tribe of South Africa thought these lights were witchfires or "baloyi" and were convinced they were sent to terrorize people who erred. The Valenge believed their sorcerers flew by night in the form of fireballs over trees. The German name for the mysterious wandering light, or will o'the wisp, is "Irrlicht". It was thought to be an omen signifying death, or a lost soul moving upon the earth with an invisible and ghostly funeral procession. (From Funk and Wagnall's Standard dictionary of Folklore, Mythology and Legend, an unabridged version 1984, 1236 pp.)

Then after coming of science these lights tended to either be ignored by scientists or were studied as a separate phenomenon, called ball lightning. Amateur investigators from all over the world were not content with the official silence and so formed there own research communities. These later observers continue to use the "UFO" label for what stands for a very old phenomenon. The "UFO" designation has unfortunately become connected with alien spacecraft-especially after the 1947 Mt Rainier sighting of Arnold. However what several witnesses often observe is an unexplained fiery orb capable of performing intricate aerial maneuvers, divide and recombine and even dig trenches.

The intrepid explorer Christopher Columbus saw a fireball a day before he glimpsed America on the 11th October 1492. It was said to resemble a large candle flame that moved vertically up and down. Jacques Vallee, a scientist, stated that behind the UFO phenomenon there is a real mystery challenging scientists:

Some people claim that UFOs are flying saucers piloted by aliens. There is even the notion bandied around in the early part of the twentieth century that UFOs could fly from a hollow earth. One extreme UFO position is the grid theory of Bruce Cathie. Bruce Cathie a New Zealand pilot saw a cigar-shaped object over Kaipara Harbor. Cathie was so captivated by this and other UFO sightings that he came up with a new UFO theory that grabbed public attention at the time. He claimed, in two books, *Harmonic 33* and *Harmonic 695,* the existence of a strange ethereal grid that flying saucers pilots used for navigation purposes. Cathie thought that the theory could predict when a flying saucer might appear at a given location. This unusual and fantastic UFO numerical grid scheme has not gained wide acceptance by scientists.

As I have said earlier, UFOs (in the extraterrestrial sense) became very popular following the media frenzy in 1947. A pilot, named Kenneth Arnold, saw (actually it temporarily blinded him) a formation of nine shiny discs around the slopes of Mt Rainier. He was searching for a lost plane. Each object was estimated to be about 30 meters in diameter. Contrary to a common perception, Arnold did not see or describe the UFOs as flying saucers. The flying saucer label came from a journalist because Arnold described the motion as being like a saucer that skipped across the water. However, the shape was *"flat like a pie pan and somewhat bat-shaped"*. This is not like a flying saucer at all. The motion of the UFOs captivated Arnold, which was *"like a fish flipping in the sun."* or *"like the tail of a Chinese kite."* A newspaper journalist Bill Bequette on June 26th 1947 wrote the following account:

"He was surprised at the way they twisted just above the higher peaks, almost appearing to be threading their way along the mountain ridge line."

The "flying saucer" label originated from a newspaper journalist who reported the story and gave a new twist to phenomenon. Since science has failed to come up with a reasonable explanation of what many observers report the controversy has raged on. Fireballs continue to spin, hover, and race across the night sky. They even terrify and cause bodily harm to people. UFO researchers who have studied these reports have called for a thorough scientific investigation into the phenomenon but mostly on deaf ears.

From the many reports of UFOs, it is clear that a small percentage were completely unaccounted for. Many UFO researchers are adamant these UFOs are not any of the common meteorological objects commonly proposed by scientists. Ufologists have claimed that the US Government tried to cover up the real story behind the phenomenon. Such a claim has served to create a division between the scientific community and Ufologists. Official scientific investigations have failed to silence the ET (Extraterrestrial hypothesis) proponents. Many UFO investigators believe that the scientific establishment tried to cover up the truth. Hence the apparent cloak of secrecy surrounding the US military Area 51, and Roswell, where it was claimed that an actual alien spacecraft crash-landed. The real answer is that the Government scientific advisors were unable to come up with a credible answer to the problem based on contemporary science. The Government has been in an embarrassing position. Though there were individual debunkers, like Menzel and Klass, who made the headlines, their scientific theories deficient. The Government sponsored a few official investigations that delved into the problem over a long period of time but with no success. These UFO inquiries raised more questions than answers.

I will briefly review the debate on UFOs, and then take a look at a selection of the many UFO theories available. I will consider those theories that treat UFOs as an earth-bound natural phenomenon rather than sociological science theories based on the extraterrestrial hypothesis, or Jung's psychological thesis.

US GOVERNMENT INVESTIGATIONS

All of the well-known major investigations, as far as I am aware, took place in the United States. In 1966, for example, a large-scale investigation, called the *"Colorado Project"*, or *"Condon Report"* was undertaken by academics at Colorado University. This investigation was the third official inquiry in a line of predecessors beginning with *"Project Sign"* which began on 22 January in 1948. In August of the following year *"Project Grudge"* was undertaken (Buttlar, 1979). *"Project Blue Book"* also an Air Force report, followed the Colorado Project and continued until 1969.

The Condon report was different than the earlier US Air Force investigations because it involved a wide range of competent professionals including astronomers, physicists and psychologists. Dr Edward Condon, a professor of physics headed the project. The National Academy of Sciences independently assessed the results of the investigation. Funding was provided by the US Air Force. Finally, in 1969, a 1465 page report was completed entitled *"Scientific Study of Unidentified Flying Objects"*. Dr Condon in his summing up of the study, stated,

"Our general conclusion is that nothing has come from the study of UFOs in the past 21 years that has added to scientific knowledge".

The report further speculated that,

"An explanation of those sightings not adequately explained will eventually be found in a greater understanding of a little understood phenomenon."

161

In my opinion the above statement of "little understood phenomena" clearly conflicts with the report's conclusion that UFO reports contain no scientific knowledge. Surely by attempting to understand such "little understood phenomena" many of these UFO reports could potentially contain useful observations that would indeed add to our scientific knowledge.

Unfortunately the National Academy of Sciences endorsed the study's findings. There was official opposition to Condon's conclusion that emanated from a prestigious scientific body called "The American Institute for Aeronautics and Astronautics". This body criticized Condon's conclusion by pointing out that a minimum of thirty percent of UFO events could not be accounted for. They said that Condon's claim that a more detailed inquiry of UFOs was of no use was not substantiated by the facts (Buttlar, 1979). Given that Condon's conclusion had been challenged by another independent scientific body of high scientific standing, demonstrates that the Condon Report was certainly not the final word on the subject of UFOs.

The Project Blue Book does have some interesting UFO observations. Some of these reports show a type of UFO consistent with twisters burning a gas inside the funnel. Don Berliner has compiled over 500 unexplained UFO sightings of Project Blue Book on his website at: http://www.ibmpcug.co.uk/~irdial/bluebook.htm.

Typical reports suggest the vortex fireball theory may apply in the following UFO sightings. There are many other like them. I have chosen a few to illustrate correspondence with the theory.

"On the 10th of February 10, 1955; Bethesda, Maryland. 10:03 p.m. Witness: E.J. Stein, model maker at U.S. Navy ship design facility. One object, shaped like a small portion of the bottom of the Moon, with a radiant yellow color, hovered for 30 seconds. Its bottom changed to a funnel shape. Total sighting lasted 1.5-2 minutes."

Quite a significant proportion of unexplained UFO sightings describe cylinders and tubes and sometimes a funnel is seen to be rotating or making a noise like bees. These features are associated with a classic tornado funnel.

"Aug. 4, 1950; approx. 100 mi. SE of New York City (39' 35' N., 72' 24.5' W.). 10 a.m. EDT. Witnesses: Master Nils Lewring, Chief Mate Jacob Koelwyn, Third Mate, of M/V Marcala. One 10' cylindrical object at 50-100' altitude, flying with a churning or rotary motion, accelerated at end of 15 second sighting."

"On Oct. 23, 1963; Meridian, Idaho. 8:35 p.m. Witnesses: several unnamed students, including Gordon. One object shaped like a circle from below and like a football from the side, hovered low over the observers, making a deep, pulsating, loud, extremely irritating sound, for 6 minutes."

"May 29, 1952; San Antonio, Texas. 7 p.m. Witness: USAF pilot Maj. D.W. Feuerstein, on ground. One bright tubular object tilted from horizontal to vertical for 8 minutes, then slowly returned to horizontal, again tilted vertical, accelerated, appeared to lengthen and turned red. The entire sighting lasted 14 minutes."

The Idaho UFO sighting is definitely consistent with a cylindrical funnel of a rotating twister which is undergoing vortex-breakdown where there is a sideways expansion of the core often giving the appearance of a bulbous shape that can also look spherical or elliptical (oval). In fact many of the unexplained cases in Project Blue Book are discs (parts of cylinders), cylinders, spheres and elliptical shapes- all geometries consistent with the three dimensional forms of a twister.

A physicist's view

Professor Paul Davies, a cosmologist, physicist, and popular writer at Macquarie University, Australia made an interesting point on the relationship between ball lightning and UFOs in a letter to *Nature* that I fully agree with. He clearly identified a tendency within the scientific community to sweep under the carpet certain embarrassing enigmas of science. In his opinion there is a three step method of doing this (Davies, 1971).

"The philosophy of this approach seems to be that if a naturally occurring phenomenon is hard to account for conventionally (i) decide that it has no physical reality; (ii) construct a physiological or psychological explanation; (ii) ignore the physical evidence that contradicts this explanation."

He suggested that this policy of denial not only applied to ball lightning but to the phenomenon of UFOs.

"All this bears a striking resemblance to the mistreatment of that other perennial and highly disreputable subject of unidentified flying objects, to which the above philosophy has been consistently applied for 25 yr. The final irony is that a section of the Condon report on unidentified flying objects is devoted to an explanation of these phenomena in terms of ball

lightning by Altschuler."

Martin Altschuler published his own ball theory and was an author of Chapter 7 of the Condon Report entitled *"Atmospheric Electricity and Plasma Interpretations of UFOs"*. I have read this report on 5/12/2003 and it is clear now that the purpose of the project was a collective effort from a wide representation of the academic world (including Bernard Vonnegut). There a range of alternative propositions to explain this type of UFO. It was like a drag net covering any known atmospheric phenomenon that might explain these UFOs. From my point of view, it is clear, in retrospect, that there was lack of recognition that some named phenomenon were of the same nature. In particular lights in tornadoes, ignis fatuus, ball lightning, strange meteors, earthquake lights and mountain electrical effects were treated as separate phenomena under separate headings. The vortex fireball theory would unite these neatly under one phenomenon.

It was not surprising that the Condon report came to such a conclusion as it did. To be fair, the authors worked hard to uncover the true nature of these lights but without success. The unified theory of these lights makes it easy to see, in hindsight, what the true solution behind the type of UFO causing all the fuss was. It is now clear that Project Bluebook contained a number of sightings consistent with combustion in vortex-breakdown.

THEORIES OF UFOs

I will examine some frequently used explanations to account for UFOs. It is obvious that a UFO could be any unidentified flying object. But what I am concerned about are scientific theories that attempt to account for highly unusual anomalous lights and fireballs.

Meteors

One popular theory, a favorite of astronomers, is to explain away a UFO sighting by incorrectly identifying the UFO with a meteor. Even a brief examination of the properties of a meteor would quickly eliminate this as a candidate explanation for many UFO events.

What are the typical features of a meteor sighting? Meteors are lumps of rock, sometimes with a large percentage composition of iron and or nickel (though there are stony meteors), that descend into the earth's atmosphere as a fireball with a tail. On entry, the meteor typically has a characteristic sonic boom, and the tail points in a direction more or less opposite to the direction of motion. In addition, the fireball is present only when the meteor is traveling at sufficient speed (i.e. supersonic), so as to cause substantial frictional impact and heat generation, as the meteor enters the earth's gaseous atmosphere. The same basic physics applies to a spacecraft entering the earth's atmosphere. This is why a shallow angle of re-entry is required as well as high temperature resistant ceramic tiles. The trajectory of a meteor usually takes place in a predictable parabolic arc. Sometimes there are changes in direction and there are cases on record where a meteor fireball has broken up into fragments as it disintegrated but even these show ballistic motion. Meteors also have small-angled trajectories

There are other significant differences between this type of UFO and a meteor. UFO speeds are known to be much less than sonic (at ground level around 330ms⁻¹). They have diverse motions with rapid changes in direction over a fraction of a second. The UFO tail, if it has one, is not normally in the direction of motion, as in a meteor's tail, but is usually vertically below it. Meteors break up into fragments in an irreversible way and could not reassemble again as UFOs have been known to do. In some cases UFOs repeatedly break up and recombine.

Optical illusion theories of UFOs

Just as optical illusion theories have been used to explain ball lightning (See Part I) it would appear that a similar explanation has been used to explain UFOs. Professor Grossman, a specialist in endocrinology at St. Bartholomews Hospital, suggested that a lot of UFOs are comprehensible if the so called "auto-kinetic effect" is invoked. The auto-kinetic effect is an optical illusion that makes you think that a stationary light source is moving around. Grossman said that the auto-kinetic effect of the eyes is enhanced if there is no fixed reference frame (Sweeny and Beaumont, 1996). But there is problem with this approach. Many UFO sightings have taken place on clear nights where the star constellations would form a natural "fixed" reference frame. Therefore having such a reference frame would act against the idea of this enhancement effect taking place.

But there are other more serious problems with such UFO illusion theories. The reported properties of UFOs are not at all consistent with the idea of an observer merely seeing an optical trick. Many observers report that unexplained luminous lights move in ways that purely illusory images could not. UFOs exhibit diverse paths and can split into several balls and then recombine. UFOs are also known to interact with their surroundings like burning and doing mechanical damage to material objects such as trees.

The optical illusion hypothesis is also vulnerable to the same kind of criticism that Charman applied to physiological theories of ball lightning. This lack of correspondence with the physical descriptions of UFOs appears to rule out this as a credible hypothesis.

Earthquake lights

Earthquake lights are unexplained luminous forms seen in conjunction with earthquake activity. The phenomenon is widely recognized as a real scientific problem without a definitive explanation. Yet earthquake lights have been used as an explanation of UFOs. Observers describe these lights as sheets of flame, columns of fire, and moving fireballs. Earthquake lights are most likely either wholly flames produced from the combustion of natural gas diffusing from the Earth's surface, as Gold (1987) has suggested, or, as I think, a combination of the usual flame combustion and vortices burning this natural gas. The vortex hypothesis has the advantage that it may explain the more unusual and dynamic behavior of the fireballs in the vicinity of the earthquake zones.

Gold (1987) page 55, suggested that all earthquake luminous activity is probably derived from simple gaseous combustion and so is of a non-electrical origin. This view is in opposition to several electrical theories and not consistent with the sightings reported by Galli (1911) who amassed 148 earthquake events where luminous phenomena were observed, many of which involved flames issuing from the ground.

But Gold's combustion explanation, as it stands, is not entirely satisfying. He was not so concerned with the nature of these lights but in providing tangible evidence for his hypothesis concerning the earth as a vast reservoir of hydrocarbons. Nevertheless, there is a still a need to explain some earthquake lights which behave like ball lightning. They seem to dart around locations where there is strong seismic activity. Barry (1966) believed that the luminous phenomena, cited in Terada (1931), and associated with earthquakes, include cases of ball lightning. Barry considered some luminous earthquake lights to be very similar to ball lightning observations. Perhaps the vortex burner hypothesis can explain some of these earthquake lights, especially those not covered by the simple combustion of natural gas.

What do observers see when they encounter earthquake lights? In 1957, after an earth tremor centered near Carnwood Forest, in Leicestershire, England, many saw "tadpole-shaped" lights flying in the sky (Brooksmith, 1980). The city of Kobe in Japan was a site of similar unexplained lights. Kobe is located at the intersection of three plates-a source of frequent earthquake activity and strange lights, including fireballs. Such tectonic activity is not surprising and appears to be in accord with standard plate tectonic theory. Musya (1931) collected at least 1500 accounts of lights from a single earthquake on the Idu Peninsula on the 26 November 1930. Interestingly the observations were both before and after the earthquake. The lights included lights like: sheet lightning, streamers, beams, fireballs and rows of round forms of light.

"The lights were very strong-in one place brighter than moonlight. They were usually described as bluish in color, but sometimes reddish yellow, yellow, or reddish blue. In shape they resembled the rays of the rising sun, searchlights and fireballs. The duration of each light was stronger than that of lightning, and some careful observers report that the same light continued more than a minute. The direction in which they pointed usually but not always to the epicenter of the earthquake."

The sustained luminosity of one light which continued for more than a minute with a pronounced blue color in an active earthquake zone is consistent with combustion of natural gas burning with a blue flame.

Some would suggest that earthquake light displays are a massive St Elmo's fire effect. St Elmo's fire can give the blue color but it would be hard pressed to explain fireballs that rove around unattached to any conductor. Somehow the St Elmo's effect has to create a very high electric field in the atmosphere, sometimes well above the earthquake. The combustion hypothesis is simpler and more plausible and appears to be consistent with known observations especially given the very real possibility that natural gas emission could have taken place during the Idso Peninsula earthquake.

Alternative explanations of earthquake lights and its controversy appeared in a book edited by Spenser and Evans (1989). In the book called *"Phenomenon-Forty Years of Flying Saucers"* some investigators claimed a link between earthquake lights and UFOs. Michael Persinger, a Canadian psychologist, claimed he had discovered a positive, though not a direct correlation, between UFO sightings and earthquake activity. However the reliability of the data that he chose as a basis of his statistical analyses has been questioned by Rutkowski (1988). Persinger plotted the locations of UFOs and seismic events on a map and discovered a general trend where the highest frequency of UFO sightings was found along earthquake zones. His generalization is supported by other researchers, such as F. Lagarde, who found a similar correlation between UFO incidence and fault lines in the 1954 UFO sightings he studied. Lopez also reported a similar correlation for the 1968 and 1969 UFO sightings he studied (Brooksmith, 1980).

The above correlations expected. The vortex burner theory predicts that vortex fireballs would tend to occur along earthquake fault lines simply because that is where vortices, fuel gas and an ignition source (eg from volcanoes) are located to power the vortex burner fireballs. Vortices are also favored along mountain ridges coinciding with fault lines.

Another hypothesis with a geological flavor to explain earthquake lights was proposed by Persinger and called the "Tectonic Strain Theory" (T.S.T.). This theory proposed that UFOs represent the visible radiation emitted by rocks under pressure in earthquake zones (Rutkowski, 1988). This idea appears similar to plasmon conception of Ohtsuki which I described in Part I of this book.

Paul Devereaux had a similar theory to Persinger's TST theory. He was a strong advocate of some kind of piezo-electric discharge effect taking place in certain rocks that could generate this effect. The piezo-electric effect is well known in physics where substances' like quartz, can generate an electric field when external pressure is applied. Devereaux used experimental evidence to support this piezo-electric explanation of earthquake lights. He mentioned Dr Brian Brady who demonstrated the existence of tiny lights seen darting among rocks in a rock crushing chamber. These lights were likened to UFOs.

Although Devereaux, McCartney, and Merron could produce a similar outcome to Brady's result, using granite, they were unhappy with the explanatory ability of the piezo-electric theory, presumably because their experiments demonstrated the presence of experimental lights in ordinary rocks. The nature of these lights was not thought to be plasma-based because of an absence of microwave emission. On the other hand spectrographic analysis of the lights revealed that the lights were dependent on the gaseous or liquid medium rather than being due to the rock itself. No definitive explanation of the lights was given. Devereaux then advanced the idea of some kind of electromagnetic radiation emitted from earthquakes.

Ball lightning theories of UFOs

While some investigators have treated ball lightning and the classic UFO sighting as two completely distinct puzzles to be understood, a few theorists have sought the answer to UFOs in ball lightning.

Acceptance of this avenue of investigation has not been without it critics. Amateur UFO researchers, in particular, have tried to attack the ball lightning hypothesis with comments based on the older perception of ball lightning as a very rare phenomenon only seen during a thunderstorm. In this view ball lightning is a small object, typically around 20 centimeters in diameter. This view is less credible given the new, more dynamic picture emerging from large numbers of ball lightning sightings where diameter can be several meters.

Scientists are also familiar with attempts to associate ball lightning with the question of UFOs. One ball lightning specialist concluded in 1997 that although issues surrounding the conundrums of ball lightning and UFOs have definite similarities they are not of the same nature. Singer (1977) wrote:

The occurrence of ball lightning has been used to explain a number of the observations referred to as unidentified flying objects. The existence question, as indeed all the information concerning the latter, is in strong popular dispute. Much of the controversy concerns questions quite similar to those which have been debated for almost two centuries on ball lightning, and it appears that both sides are unacquainted with the existence of this precedent. Many of the of the questions are familiar to those who have considered the problem of ball lightning, as are the results of the debate. The similarity of the two subjects in this respect do not extend to the nature of the observations or the information they provide. A definitive discussion of this issue which concludes that unidentified flying objects do not exist has been presented by Jones (1968).

Singer seems to be suggesting that although the debate and its history have been similar the observation details pertaining to the nature are different. At that time of his writing the properties and descriptions of ball lightning were somewhat narrow-this has changed and the ball lightning-UFO connection is far more viable than at that time. There are, indeed, details that are common between the two subjects of inquiry.

One reference to ball lightning as explaining a type of UFO, (although the UFO label did not arise until around 1949) is contained in Timothy Good's book "*Above Top Secret-The World Wide Cover Up*". It mentioned a newspaper report which proposed various suggestions to explain away the so called "foo fighters". The article was published around 1943-1944. The United States 8th Army wanted a searching investigation into the foo ball phenomenon. However no satisfactory answer emerged. Various explanations were proposed at the time including, St Elmo's fire, ball lightning, and combat fatigue. But these explanations were thought unlikely to account for all cases. (Good, 1989).

Klass's ball lightning theory.

Philip Klass used a controversial ball lightning explanation in the form of corona (St Elmo's fire) to explain UFOs. In a publication entitled "*Aviation and Space Technology*" he postulated (Klass, 1966) that many UFO sightings could be explained simply as an ionized ball of gas, or corona, created by high tension wires. Klass's theory on ball lightning unfortunately has serious limitations. His theory is limited because it is not capable of fully explaining many of the reported properties of ball lightning. In his corona theory ball lightning has to be attached to a high voltage power line capable of generating corona. The aerial corona theory also has no obvious explanation for the diverse shapes and motion attributed to ball lightning.

Tambling (1967) criticized Klass's theory on the basis that the corona theory depends upon the presence of high voltage power lines in the vicinity of a UFO. Indeed Tambling referred to Mr Ferriere and Mr Trudel of Woonsocket, Rhode Island, USA who both saw UFOs near power lines that were not the type of high voltage lines needed to create corona. So where is the corona going to come from in these fair weather conditions?

Another observation in favor of the vortex burner theory is the following observation made by Mr Ferriere; editor of a UFO magazine called "*Probe*" and quoted by Tambling in his book. Ferriere said that the marsh gas (i.e. mainly methane) was very active the afternoon of July 24th 1966 of the sighting along Diamond Hill Road. The summer conditions and hotter temperatures could have contributed to an increased methane emission from the swamp. Research has shown that temperature increases are positively correlated with increases in atmospheric emissions of methane from swamps and bog lands. One study found that when there was a temperature rise from 5 to 15 degrees Celsius there was a hundred-fold increase in the methane concentration (Krill et al, 1988). So hotter weather could explain why the marsh gas was activated. While Klass's ball lightning-corona theory, when applied to these "marsh gas" UFOs, is clearly unsatisfactory. However the vortex burner theory is consistent with the existence of natural gas in the swamp zone and the beauty of such an explanation is that it does not require high voltage power lines for a fireball's existence.

One apparent weakness in Klass's theory was pointed out by another UFO researchers, Jenny Randles. She was a critic of ball lightning as a suitable explanation of many UFO sightings (Randles, 1987). In "*The UFO Conspiracy -The First Forty Years*" she criticized the idea that ball lightning could be used to explain certain UFO sightings. Her view of ball lightning appeared was that of a very rare and unusual form of globular lightning- certainly not as large as the width of a road or long lasting or having any effect on the engines and lights of cars. Unfortunately her comments revealed several assumptions inconsistent with modern research on ball lightning. One of them is that ball lightning is always associated with lightning strikes. Ohtsuki has reported field evidence of large numbers of fair-weather ball lightning sightings. This new research alters the traditional picture of ball lightning as something seen during a thunderstorm. The other misconception is that ball lightning is a very rare event. Several surveys have demonstrated a diverse range in both size and lifetime.

Rutkowski (1988) another critic of Klass and a UFO investigator, mentioned an alternative proposal to the corona theory of ball lightning. He wrote that Klass had claimed that there is evidence that many UFOs can be accounted for as a form of ball lightning electromagnetic plasma. Rutkowski was skeptical of ball lightning theories of UFOs. His assumption that ball lightning is very rare, as I have said before, is no longer viable. Ball lightning is far more common than since recent research has revealed that more and more cases of ball lightning are being reported and collected each year.

Tore Wessel Berg's theory

Individual scientists continue to believe there is a connection between ball lightning and anomalous lights. This ball lightning-UFO connection is not new and follows a path followed by Klass and others (including me). The real crux is to get a successful theory of the ball lightning problem. Tore Wessel-Berg (2003) proposed his theory for ball lightning then went on to more explicit associate his theory with UFOs in a paper to the *Journal of Scientific Exploration* in 2004. The theory is centered on a standing wave of direct current pulses in air from an electrode-less discharge. This is a direct application of Maxwell's electro-magnetic equations. The theory requires a direct lightning discharge and there will be a decay of the energy in a region outside the less energetic ionized zone. The most obvious shortcoming is the requirement not only of a lightning discharge but one where additional initialized conditions are required. There are other properties that need to be explained including the mechanical ability of ball lightning to dig ditches and drill holes and to recombine and combines again.

Comments on the ball lightning-UFO connection

Not all UFO authors assume that ball lightning is a rare and exotic phenomenon and could not possibly explain the intractable UFO cases often reported. Cade and Davis (1967) in their book, *"The Taming of the Thunderbolts"* discussed the myths and science behind ball lightning available to them at that time. They supposed that since a fraction of UFOs are luminous spheres and since ball lightning is commonly seen as such a glowing sphere, then perhaps some UFOs are ball lightning. Page 105 of their book suggested that ball lightning may be a suitable explanation of these flying saucers. Unfortunately a satisfactory ball lightning theory to explain these UFOs was not available to Cade and Davis. This is quite understandable given the plethora of competing ball lightning theories.

It is evident to me that several UFO researchers have not realized the diverse range of the physical properties of ball lightning from the many published observations that are now available.

Meaden's crop circle theory.

Another explanation of UFOs is centered on the so-called, "circles effect", put forward by a meteorologist called Terence Meaden. The circles effect is a phenomenon relating to the flattening of crops, such as wheat. The flattening is usually contained within a circular shape, though other more complex unlikely geometrical arrangements have been documented. Several investigators have put forward different explanations as to why these crop circles have come about, from hoaxers with tools to flatten the crops, to aliens and flying saucers (i.e. the extra-terrestrial hypothesis). Unfortunately the credibility of the circles effect as a real scientific phenomenon has been somewhat damaged by the incidence of hoaxes. A number of scientists went to England to investigate these crop circles. Among these were Ohtsuki from Japan and John Snow well known in the field of columnar vortices.

Meaden's scientific explanation of the circles effect was of interest to me. In a book entitled "The *Circles Effect and Its Mysteries*", Meaden (1989) postulated that the circles effect could be explained by a type of whirlwind or vortex illuminated by an electrical discharge.

Meaden's attempt to unravel the mysteries of the crop circles resulted in a book which he self published. The logic of the basic argument behind his approach, in its barest form, seems to me, to be this. The crop circles appear as though they have been flattened by a whirlwind. Since some observations of balls of light have been seen near these flattened circles, perhaps the balls of light are vortices emitting light resulting from electrical activity.

His aim was not to formulate a theory of ball lightning, but to solve the riddle behind the crop circles. His view of ball lightning appears to be based on a popular misconception about ball lightning's properties. The definition of ball lightning in his glossary defines ball lightning as a small spherical ball around 0.4-0.5m in diameter. The ball is short-lived (under 5 seconds) and it's electrical nature is unknown. He assumed ball lightning is fundamentally an electrical phenomenon, which he thought to be a cool plasma object where the energy is stored in metastable molecular energy levels within atoms of molecules such as oxygen and nitrogen present in the atmosphere. He also assumed a number of physical aspects about ball lightning which need to be revised in the light of the current (even in 1989) state of ball lightning research. For instance the lifetime of ball lightning can be long-lived, and fair weather ball lightning is quite common in some parts of the world eg Japan. The assertion that ball lightning is of an "electrical constitution" has not been validated.

His description of ball lightning, having "metastable molecularly-excited levels", probably originated from Powell and Finklestein (1970). This was a ball lightning theory based on the electroluminescence of air. Apparently Meaden regarded this cool plasma theory as the best current explanation of short-lived ball lightning. However this theory has not been universally accepted as the standard explanation by those in the field.

At another place in his book, Meaden went on to differentiate between the traditional form of ball lightning, explained using electroluminescence, and another longer-lived phenomenon and that using the recombination of electric charge along the lines of the Powell and Finkelstein mechanism, but with the electrical activity in a vortex. He termed this longer lived phenomenon, called "balls of light" or "BOLs", for short.

His conception of ball lightning and its relation to his so-called "plasma vortex" explanation of the circles effect is inconsistent and confusing. Meaden suggested a distinction between ball lightning and the phenomenon to be the cause of the circles effect, which he named the "plasma vortex". The plasma vortex is a common theme in ball lightning and he seems to have suggested that plasma vortices and ball lightning are closely related but not necessarily identical in nature. This could be because of his more narrow definition of ball lightning. He considered the plasma in the vortex to be a result of a cool discharge (i.e. low current discharge) along the lines of Powell and Finklestein, though not necessarily identical to plasma but involving metastable states. Incidentally the idea of an electrical plasma vortex as an explanation for ball lightning has been postulated several years ago (see the review of Singer, 1971).

Meaden conceived of these plasma vortices as being made luminous by the following mechanism, described on page 75 of his book. He suggested that the light generated in the postulated circles effect originated from a low density plasma vortex spinning at a high rate. This was said to take place at high altitudes involving ions and electrons. A simple recombination of electric charge was said to be responsible for the electric discharge with accompanying light and sound generated. The vortex becomes greatly charged because of these spinning electrons and ions and, it was added, there is an enhancement of the fair-weather electric field above that of $100Vm^{-1}$. Presumably what he means here is the anomalously high electric field associated with the atmosphere rather than the vortex itself. As far as I am aware such large increases in this electric field are normally linked to stormy conditions or through dust clouds which can have high electric fields.

A secondary electrical effect was mentioned where the vortex touches down on a dusty ground. Here the dust is dragged from the ground into the vortex. The dragging process would tribocharge the dust so that the dust becomes charged. This assembly of dust particles would create an electric field. I demonstrated this "touch down" tribo-charging of dust and powder in a dust devil vortex (Coleman, 1990). Significant charging coincided with dust lift off a glass floor in the vortex generator. I have not come across any work in the scientific literature on small scale dust devil vortices which actually demonstrated this charge separation process. It could be that this tribo-charging charging effect is responsible for the high electric fields of dust devils vortices that have been measured in the field by

Freir (1960) and Crozier (1964).

Several questions come to mind in the above conception. If high vortex rotation rates are required at high altitudes to generate the required light and sound, how is it then possible for such objects to exist at ground level or at medium altitudes? If the vortex happens to be at ground level why the high altitude mechanism does suddenly stop and the tribocharging mechanism suddenly take effect? Also there seems to be no mechanism to account for medium altitude luminous vortices.

Meaden's proposal of electrical vortices, (which I read about in 1996), had some features that were broadly similar to a conception of ball lightning I had been working on while I was doing my thesis around 1989. But I was not satisfied with thought experiments; I wanted to test the idea experimentally to see if a viable mechanism could generate light by electrical means within a vortex. At that time I had the conception of an electric recombination in a vortex and speculated on the possibility that the intake of electrical charge up the vortex could serve to increase the longevity of the luminosity. However I abandoned this approach once I grasped that chemical combustion was a more simple and versatile idea and better fitted the observations of ball lightning.

There was also a difference in overall approach in that Meaden sought to explain UFOs associated with the circles effect while I concentrated on ball lightning and saw a suitable ball lightning theory as an avenue which could be applied to explain a certain kind of UFO. The advantage of the vortex burner hypothesis is that there is no requirement for complicated arrangements of charge because the fuel gas can be directly supplied to the vortex. So long as the fuel gas is supplied the luminosity can continue. There is also no need for high spin rates or dramatic distortions of the local electric field etc.

Meaden added a further feature to explain more complex vortex luminosity shapes. He speculated on the existence of intense currents which he termed an "ion race". This "race" was generated by electromagnetic induction and served to feed ions into "satellite" crop circles. Four crop circle satellites are connected to a large larger circle. Together they formed a five, like that on a dice. Meaden also invoked a Kapitsa-type electro-magnetic wave mechanism to concentrate the ions in the outer satellites. But the intense currents and electro-magnetic induction seem a radical departure from the original mechanism he advocated-the low current discharge of Powell and Finklestein. Meaden speculated that the UFO that made the crop circles was a type of plasma vortex which generated its luminosity by electrical means through an energetic spin regime. A cool plasma, along the lines of Powell and Finkelstein's theory, was invoked. Meaden's theory appears to have problems of consistency and the mechanism seems confusing and unrealistic.

To conclude this chapter: The scientifically-inspired UFO theories I have described here have difficult problems to overcome in certain aspects. In my opinion there is clearly a need for a better explanation of the UFO phenomenon. In the next chapter I intend to illustrate how the vortex fireball theory is better able to account for a dynamic class of UFOs. These are the UFOs that Arnold saw, plus the classic Kaikoura UFOs, the will o'the wisp, foo fighters and many other sightings. This unified theory could apply to other fireball events that are recorded in the literature.

8

The Mystery of UFOs Explained.

"Thousands of sightings, many reported by persons of complete authority and reliability have proved beyond doubt the physical nature of the phenomenon. It is now upon science to solve the problem of their origin and to take a new look at the saucer mystery."

[From the book *UFOs Over the Southern Hemisphere* by Michael Hervey]

Introduction

Some UFO case studies demonstrate that a large proportion of these lights can be explained using obvious explanations such as: lights on aircraft, Venus etc. While some are hoaxes, such as a Ford Falcon silver hubcap thrown in the air, less than 10 % of all sightings remain unexplained. These residual cases cannot be ignored. The challenge is to find an explanation for these events. Typically these UFOs manifest a unique flying ability, sometimes involving highly acrobatic aerial maneuvers. We are not talking here of a passive object, like St Elmo's fire, or some other luminous entity.

The famous episode of the Kaikoura UFOs is classic case in point. During the 1978-79 period, when they were seen, a strange ominous feeling spread across the length of New Zealand. These UFOs were not merely stationary objects as TV film footage displayed dynamic movements including complete loops in the sky. The armchair critics speculated that it could have been Venus rising, optical refraction, or mere camera shake but these explanations were not sufficient in themselves to convince first hand observers. Clearly the people who saw this UFO were not naive observers. Captain Startup who flew the Argosy at the time of these UFO had already clocked up many years of flying experience.

Without a successful scientific explanation of these lights a large following has developed that actively promoted an alien encounter scenario. Are we really being visited by aliens from outer space? Before we leap to this position Occam's razor should be used. This theoretical guide suggests that it would be folly to make things more complicated or more fanciful than what is really required. The strategy should be to go for the simplest possible answer that gives the correct answers.

UFOs are almost all seen in the Earth's lower atmosphere called the troposphere. It is entirely logical to seek a solution in the field of meteorology. Several authors have promoted meteorological ideas like lenticular clouds, Venus, sundogs, and bright stars but this list is insufficient to account for the whole UFO phenomenon. These explanations have been correct on a few occasions but inadequate in others. Many UFO observers are dissatisfied with these options.

On the other hand most of the scientific community has yet to treat UFOs as a credible field of study. There have been a few brave scientists who have attempted theories of the UFO phenomenon and a deeper appreciation of the difficulty behind the UFO mystery.

Some believers in alien UFO craft cling to their view with a bulldog-like grip. Such fantastic speculations, while entertaining, distract the honest inquirer from the true vocation to uncover the facts.

Can all UFOs be explained using contemporary science? There is a case for fully explaining UFOs within the present paradigm of science despite previous attempts. The cause of the typical unexplained fireball at the core of the controversy appears to be attributed to a twister or vortex on fire. This could be the sought-after answer to the mystery of the classic UFO controversy. There circumstantial evidence for this proposal. A significant number of UFO sightings show signs indicative of a fiery vortex. Scientists, until now, have not understood or conceived of the existence of this type of atmospheric vortex and they have had to rely on existing outdated explanations that just do not fit the reports. Small or large atmospheric vortices, it is only a matter of scale. Most UFO sightings take place at night and such reports demonstrate distinctive properties that point towards a spinning vortex suspended in air and burning a combustible fuel gas to produce a brilliant flame-like structure. Under this theory the flame is what the observer mostly sees. But a large spherical flame that moves in the dead of night would present an awesome sight.

In 1999 my UFO theory was reported in the popular media in New Zealand. The Christchurch Press published the experimental setup at Canterbury University and a February issue of Wellington's Dominion newspaper published accounts of the theory with a front page color photograph of an experimental twister on fire. The theory was aired on the TVNZ Channel 1 news and interviewed on a local TV channel, CTV. News of the UFO theory was a published in the 22 May 1999 issue of New Scientist. A film of the experiment that produced small burning twisters was incorporated in a documentary called 'Ball Lightning'. These documentaries ran for several sessions on National Geographic and Discovery channels in 2001. In this documentary I was able to demonstrate gaseous combustion in a typical vortex with compressed natural gas (CNG) and liquefied petroleum gas (LPG) in a combustion laboratory at the Chemical and Engineering Department at the University of Canterbury.

A candidate UFO theory needs to be in accord with the evidence.

There is a sufficient body of observations from within the scientific and popular literature that circumstantially supports the theory that UFOs exhibit vortex combustion properties. In the last chapter of this book I will take a look at a small number of UFO events that appear consistent with the vortex fireball theory. The central idea here is that a small percentage of these vortex fireball UFOs are the root cause of the UFO activity that observers all around the world have got so excited about. Furthermore, there is little doubt concerning the existence of such twisters in the Earth's atmosphere, though these type of fiery twisters reported in the meteorological literature have never been well understood nor explained. It could be that this type of UFO is actually a vortex in a state of burning. What the observer sees as a UFO is, in reality, a vortex on fire with the combustion flames confined within the vortex. Such UFOs really are 'chariots of fire'.

The theory is predictive. UFOs are more likely to be seen in close proximity to the world's fault lines, especially where there is gas emission before earthquake activity. That is why some UFO researchers like Persinger and the earthlights theorists believe there is a correlation between UFO sightings and fault lines. A few UFO theorists have speculated on the existence of strong ionizing electric fields above the fault lines where UFOs have been seen. But this may not explain these UFOs. It was speculated that high electric field strength can cause ionization and with it light. However, this hypothesized effect has never been fully established and there are serious problems with the theory such as explaining the dynamic behavior of glowing spheres moving erratically.

The fact is that many UFO sightings show tell-tale signs of a vortex being implicated: the tubes and cylinders seen, oval and ball shapes, the rotation and spiral formations and generally geometric shapes. These properties are all associated with naturally occurring vortices. This theory could become the correct explanation identifying a vortex on fire as being behind the UFO phenomenon.

Specific UFO events

I will now examine more UFO cases that are consistent with the vortex fireball theory. Obviously explanations given in an earlier chapter, for ball lightning's properties, would also apply to a type of UFO which I believe is causing all the commotion. In fact I found collections of UFO sightings containing frequent descriptions of UFOs in terms of fireballs and flaming cylinders and the like. Simple geometric shapes are described rather than sophisticated vehicles from outer space. Geometries like the sphere and cylinder are no coincidence since they are a tell-tale signature of a meteorological origin-a fireball on a twister's axis.

Take a typical UFO description that is consistent with my vortex on fire conception. A man called Parnham was driving in his car from Carnesyille to Lavonia, Georgia on the 29th June 1964 when he came across a UFO shaped like a "giant top" with a "tower-like" extension. This object was spinning at a high rate and sounded like a mass of snakes hissing. Not only that but the object left an oily residue on the car and the object smelt like embalming fluid. The object traveled along with the car and managed to burn the car's paint. There are more of these sorts of observations in the media. This one is notable because the oil film on the car is a direct indication of the type of fuel that the vortex is burning. If the UFO was from outer space and of high technology it would not using oil or leaking it! The most likely explanation is that oil came from a natural reserve.

There are not many mainstream scientists willing to examine the UFO phenomenon objectively but there are some scientists who belong to organizations that deal with fringe or anomalous aspects of science. My intention was to get their opinion on my theory. This time I wanted to test out the waters with scientists who are familiar with the UFO controversy. Bernard Haisch is an astronomer and physicist and a former editor of the *Journal of Scientific Exploration*. He was instrumental in authoring a new theory on the origin of inertia arising from the electromagnetic quantum vacuum. He also belongs to a UFO skeptic organization. I sent him the following email on 6th January 2004. He replied two days later.

Bernard Haisch
Palo Alto, California

admin@ufoskeptic.org

Dear Bernard,

*I read your piece on the website and I felt inspired to communicate my theory to explain a set of anomalous lights seen in the lower troposphere. I would like your measured opinion on the theory as an independent scientist. I have published the theory in a few scientific sources such as Weather (a Meteorological publication) and in EOS Transactions. I have put the case for the theory in a book published in 1997. My latest book **Great Balls of Fire- A unified theory of ball lightning, UFOs, Tunguska and other anomalous lights** is to be published by Universal Publishers. My proposition is that there is now a satisfactory natural phenomenon to explain a large fraction on the unexplained UFO cases. Although there is still a niche for the ETH hypothesis it is has less room. But in the spirit of Occam's razor there is no need to invoke the ETH hypothesis. This natural*

*phenomenon has not been recognized and explained by scientists until I published the hypothesis in 1993 in Weather. No one had conceived of such a solution and I have read enough of the literature in the area to realize that it was a new discovery. The concept is little known or appreciated in the science community yet it has been demonstrated in the laboratory and the work published in an issue of EOS Transactions on the American Geological Union. But there are meteorological observations that suggest that such an effect does exist but that it was not understood. The idea is simple and seems mundane but I make a case for the theory and it does explain several cases of unexplained lights. In fact in astronomy anomalous meteors have been reported that deviate from the usual bolide characteristics i.e. unusual trajectory, slow, unusual physical effects. Professor Clyde Tombaugh the discoverer of Pluto has seen such fireballs. My unified theory is that many of the lights can be accounted for by the idea of a combusting vortex in the atmosphere. I envisage not just tornado size vortices but scaling down to much smaller diameters of a few centimeters. A combustible gas like natural gas is sucked into the vortex and is burnt. Normally UFOs are seen at night so the luminosity from the combustion process is what would be normally seen by the observer. I have photographs and written accounts of UFOs that show signs of this combustion process and at locations lower down in the funnel. But how could a flame survive in such high air speeds? The answer is vortex-breakdown of a naturally occurring vortex with a substantial reduction in air speeds and associated stagnation zones. Luminous shapes of some UFO and ball lightning photographs show the distinctive vortex-vortex-breakdown signature. Most of the scientific work was published through the University of Canterbury and Lincoln University, New Zealand.
Regards Peter F Coleman MSc MAppSci*

Bernard Haisch's reply:

*Yes, this could explain some sightings, but there are a huge number of reports of structured craft, interactions with entities of some sort, etc. I'm afraid the reported phenomenon goes far beyond peculiar lights, and there is the danger that, eager to brush the whole phenomenon off, scientists will grab onto this and say, "Oh, that explains it all."
Regards,
Bernard Haisch*

The first point is the scientific basis of my fireball theory stands its ground and Dr Haisch concedes that the theory could explain some UFO sightings. He is right in saying there exists a body of literature encompassing various aspects of alien encounters including the craft itself and the reference to *"entities of some sort...."* However the UFO reports I have seen were mostly concerned with unexplained lights rather than aliens. What proportion of events classified as UFO events actually represent actual alien sightings? I suspect the figure would be small. I have yet to see such a ratio or its statistical estimate. In my view the plausible alternative is that these alien visitor reports could be an imaginative reconstruction of an unexplained natural phenomenon. This proposition cannot easily be ruled out especially given that the experience of a vortex on fire would be an awesome sight. Couple this with the fact that mainstream science has yet to officially recognize the existence of these vortex fireballs so observers have no valid theory of what they are actually observing. I should say that many of the UFO sightings I have read about manifest themselves as *"peculiar lights"* rather than as alien craft.

From a human emotional aspect these UFO lights have been interpreted as ghostly apparitions, chariots of fire and souls. Lights hovering in graveyards are a common occurrence in some cultures. In Japanese graveyards, ghost-like lights have been seen hovering over the remains of ancestors. Susumu Yamauchi who lives at Yasuda ku Tsugawa machi wrote:

> *"It was 31 years ago in September when I witnessed it...it happened around 11.30 pm. I was looking outside the window, without paying much attention, and then a lantern like light appeared. First, I thought it was a lantern, but then I looked more carefully, and noticed it was a sphere with irregular motion... The object was about 20 cm in diameter shaped like a sphere. I had thought that ball lightning moved in a straight line, but this moved irregularly, so I was scared, but at the same time interested. Before the object disappeared it flew straight up, made a circle, and then moving in the direction of the rice field disappeared. The next day, one of my students came to me and told me his grandmother had passed away. I found out later, that the direction, in which the object disappeared, was in the direction of the cemetery."*

[Quoted from an unpublished compilation of sightings presented at an International Symposium on ball lightning conference by Makoto Egawa].

Therefore, the notion that these lights were thought to be the heavenly souls (hitodama) departing the body would not have seemed odd in the least way.

Another mythical association of these lights is connected with lantern-like lights that have been seen moving down a mountain called Kanabera Mountain, opposite the town of Tsugawa. Tradition has it that these lights represent the lights carried by foxes as a wedding ceremony makes its way down the mountain. Even today, the women in the town paint their faces white in imitation of this fox wedding procession. Yauko Dohi gave this testimony:

"I witnessed a kitsunebi (mysterious fire) in the fall of 1946. It was right before sunset and getting dark. Suddenly, there was a lot of noise coming from outside. I was curious and went outside to see. Then at the base of Kanabera mountain lights the color of a lantern made a horizontal line, and repeatedly appeared and disappeared. Then lights gradually became hidden by the mountain ridge and were gone. There were many other people witnessing this with me. This phenomenon, which lasted 20 minutes, reminds me of the story my mother told me about the kitsunebi. The fairy tale came true. It was extremely mysterious". [Makoto Egawa compilation]

The Japanese popular obsession with these lights is demonstrated in a ball lightning picture on their currency and a museum on the East Coast of Honshu is dedicated to the cult of the UFO.

Many UFO reports have been published in books and circulated on the Internet on websites dedicated to the cause. These sources contain many legitimate sightings. Michael Hervey in his book, *UFOs over the Southern Hemisphere,* has a wide range of interesting sightings from Australia and New Zealand. Among these reports there were UFO sightings that hinted at the existence of twisters on fire.

...I rushed into the house and got my telescope which showed that the disc had <u>an intensely bright golden center, around which the white part rotated</u>... [Bertrand Collin from Dannevirke, New Zealand]

...about five minutes it suddenly reappeared as a bright red ball. We watched this change to an oval shape, then a <u>cigar shape</u>, all the time getting lower.... [Mrs H.K. Jukes, Invercargill]

A wave of UFO sightings was reported in 1969 in the North Island of New Zealand. Here are two sightings from that collection. The UFO's tail is circumstantial evidence for the existence of a vortex where the flame structure is modified by the swirling air mass.

....A light in the sky 'as big as a football, rather like <u>a ball of fire with a greenish tail</u>...

.... a loud <u>humming</u> multi-colored object as long as a school bus hovered over a paddock close to them...

Twisters can make a lot of noise when the observer is close to the funnel. This account is consistent with the expected sound that would be heard from an observer in close proximity to the funnel.

...A Thames woman saw a 'bright, spinning light' across the Firth of Thames..."It was round and saucer-shaped and looked like the flying saucers in "The Invaders".

Six people in Wellington saw a UFO for an hour in May 1968.

Four of the people who sighted the UFO had it under observation from 10.30 pm until 11.30 pm. They said the <u>object was shaped like an egg with both ends chopped off, was continually moving in a circle</u> and upwards and downwards and emitting sparks of varied colors"

An observer while they were waiting for a taxi at the Queanbeyan railway station at about 4.30 a.m. saw a:

"...<u>silvery object revolving in the sky with blue flames streaking from it.</u> It seemed to travel in quick circles, and with each turn the flame shot our. I was not the only one to see it-there were two other ladies present, and three members of the Air Force, also waiting transport."

These last two observations are typical UFO accounts. They are consistent with the property of vortex-breakdown where the funnel of the twister can expand laterally and create a bubble-like recirculation zone. Rotation in a circle is characteristic of vortex motion. The book *Phenomenon –Forty Years of Flying Saucers* edited by John Spenser and Hilary Evans, documented a UFO which was seen in the province of Guizhou on the 24 July 1981. The apparent diameter was about the same size as the moon but it had a luminous tail. The tail became a set of spirals around the central body of the hub. One observer likened the UFO to a Chinese dragon.

"...like a washbasin, of a diameter he estimated at about twenty meters, <u>spinning </u>and throwing out flashes of brilliant blue light, surrounded by misty white swirls of foggy white..."

Although many UFO sightings do not make it to the scientific literature, some are worthy of genuine investigation. In the book *The Evidence for the Bermuda Triangle* a couple, called the Winfields, encountered a UFO while they were on a fishing trip. The report showed evidence of burning inside a tube extending to sea. This observation of a UFO is the best I have read where the funnel of the vortex is reportedly on fire with its accompanying smoke and entirely consistent with the twister on fire theory. While other sightings point circumstantially towards the fiery vortex theory, this report is much more explicit in its reference to the main elements of the theory.

"About 6 kilometers off Boca Raton, Florida when at 2pm, Jean notices a stream of smoke along the horizon.... As they neared the source of the smoke, the more it appeared to be a ship on fire...they received a shock: the source was not a burning ship, but a pipe, 20-25 centimeters in diameter belching flames and thick smoke. They noticed the pipe was yellowish-colored, as was the smoke, and from 30 meters away produced no smell or sound. After they watched it for some time, the smoke and flames gradually subsided, leaving just the pipe protruding from the waters."

Natural gas has been reported as bubbling at the sea surface and would have provided an ideal source to feed the combustible gas into the burning vortex. Thomas Gold, the noted scientist, included this quote on page 54 of his book *Power from the Earth concerning* vapors that were released from the Earth.

"...There have been other effects upon the water, such as a surprising flux and reflux of the sea, extraordinary agitations and commotions of the waters..."

In an October 1988 issue of the Australian Women's Weekly there was an article about UFO encounters in Australia. Dr Peter Turnbridge (GP) saw a ball of light racing across the sky above the Adelaide Hills. There are no other clues in the verbal description to the object being positively identified as a twister but the photographs do indicate a tail-like structure as the following statement demonstrated:

"It was a very bright white light in the sky-a light which did not have any shape...when you slow the tape down to single frames, (1/25th s) the object has almost a plasma appearance. It looks like a ball of energy. I don't consider it to be a flying saucer, but I don't know what it was."

This UFO could obviously move fast, but it was also reported at one point to hover for a while in one place. This movement is typical of twisters that have been seen to remain suspended above the crest of a hill. In the same article, a strange green light was seen about 30 kilometers south of Marla Bore in the upper north region of South Australia.

"Lisa says, " We looked to the left of the van and there was this strange,

fluorescent green light between one and two meters across, about 10 meters off the road and a couple of meters up in the air. It was rotating, like a wheel within a wheel, but not moving along. The light was in concentric circles it was dark."

This curious green UFO shows characteristic vortex activity with the rotating air mass. The green color might be due to the presence of copper compounds in the desert region. It is well known that when a copper salt is heated in a Bunsen flame shows a distinctive green color to the flame.

The above examples from the annals of UFO observations are only a very small sample, nevertheless, on a case by case basis, they are consistent with the concept of a burning vortex at work. Furthermore, what is evident is that none of the observations report a well-defined aerial spacecraft.

These fireball vortices have been given little in the way of wide public exposure but they do appear in the meteorological records. An important question needs to be addressed. Have these type of twisters suddenly disappeared or have they been reclassified as UFOs simply because science was bereft of an explanation? Vortices are a ubiquitous feature of the Earth's atmosphere. With the ever-present vortex and the release of a combustible gas coupled with a source of ignition there is every reason to suggest that fiery vortices have been around for a long time.

Min Mins

The word "min min" is an Aborigine name for puzzling lights which have been seen for many years in certain parts of Australia. Min mins are the Aborigine equivalent of the English will o'wisp, or Japanese fox fire. The lights have a legendary status in Aboriginal culture. Bob Young, the head technician of the Electrical Engineering Department told me in 1989 that telecommunications engineers working in the outback have seen these strange lights. They look like tilley or lantern lights hanging in the air above the desert floor and they are reasonably common occurrence there.

Pettigrew's Min Min theory

The inverted mirage or 'Fata Morgana' refraction theory of Pettigrew (2003) was proposed to explain the min min lights seen in Australia especially around the area of Boulia in the channel country of Queensland. Car headlights were said to be an ideal source but campfires and the planets were other sources. Such UFO refraction theories have been casually proposed in such cases as the 1978 Kaikoura case. The visual effect could explain some types of lights but not all of them. The theory is based not on the usual mirage from warm air layers along the ground where light from parts of the sky bend downwards to follow a convex path to the observer. Instead an inverted mirage is formed by light directed along a concave path through cold stable air layers along the ground. Therefore a light source beyond the horizon generates light that bends over the horizon and on to the observer.

This stringent requirement for an inverted mirage obviously would not explain lights seen during the heat of the day or when the wind was blowing. The author has taken a narrow description of the lights and he has excluded other important features. Min mins may have other properties such as sound, dynamic interaction with the environment, the lifting of sand (the Knowles UFO), and a visible parallax effect. An Australian park ranger took a photograph of what he described as ball lightning. It did not look at all like the classic spherical fireball but a rope-like object. It could equally have been described as a min min. It was an astonishing 100 meters in diameter and having a duration of 5 minutes. The glowing object was caught on film by a park ranger Brett Porter in Queensland, Australia. This luminous object did not have the form expected of a faint car headlight. It had a long rope-like projection extending from what appears to be a hot glowing globular region contacting the ground. This color photograph was publicized at *http://news.bbc.co.uk/2/low/science/nature/1721473*.htm and also reproduced in Transactions of the Royal Society (Abrahamson, 2002). I found it could be explained with my theory and explained as follows. The rope-like shape projecting out is the twisting whirlwind core while the main luminous region at the ground level is intense combustion with the vortex-breakdown region. Here was a case where the basic seat of combustion originated from the stagnation zone within vortex-breakdown.

One problem with the Pettigrew theory is that there is no candidate light source before the advent of cars that is able to account for observations reporting real parallax motion to a human observer. Under the Pettigrew theory the only luminous sources available to the observer would be the ubiquitous moon and planets and campfires (a few hundred kilometers away) have very small parallaxes... They are essentially stationary to the observer for the short time span of a sighting. However reports of min mins do sometimes show motion effects independent of the observer and not merely small shifts associated with changes in the refractive index of the air.

Recent observations would seem to indicate that the min min light is commonplace. Fred Silcock has collected around 500 min min events in his research (Natsis and Potter, 1995). For instance, on Della Lanahan's sheep station in Queensland, a strange light was seen which had a diameter of 4.5 meters. The light reduced in size and changed to a red color. The object displayed a most remarkable behavior very similar to ball lightning. When Lanahan fired at the object, it expanded and glowed for an hour. It contracted again, went out, and then reappeared and then disappeared again.

Min mins are also well known to the locals in Townsville, Queensland. One case involved a person walking along the Townsville breakwater in the evening. The person was surprised to see a small intensely bright light, like an electric arc light, which appeared over the stony surface on top of the breakwater that moved upwards and quickly disappeared. No lightning was seen and the weather was calm. There was no suggestion that the light was electrical in nature, despite the arc welder reference, which had to do with the brightness of the light. When I first encountered this Townsville case I considered an electrical discharge originating from static electricity as a possible mechanism. I imagined a type of natural capacitor built from charged salt aerosols in the air above the breakwater. But the discharge would have to continue by some airborne electrical separation mechanism. An aerial salt capacitor did not seem a practical suggestion at that time. On reflection, the movement of this min min through the air is more likely to have been a combusting vortex lifting off the breakwater. Electrostatic discharges may be involved in igniting the gas. However, an investigation of the possibility of natural gas emission from some source near the Townsville breakwater, or a natural gas build up from a more remote source would test the combustion idea.

Tulley UFOs

Min mins are also common at a place called Tulley. Tulley is north of Townsville along the Northern Queensland Coast, not as far as Cairns. Tambling (1967) quotes a letter from Mrs Noble who lived in the Tulley area. She gave numerous accounts of an orange light descending on to a sugarcane farm, flying machines, round white lights and other strange lights. She believed the lights originate from the coast and go inland up the Tulley River valley, or Murray River and other valleys. Such lights were said to be the subject of many an Aboriginal story.

It is probably no coincidence that such lights favor Tulley. It is noted that vortices do come in from the sea and the area abounds in volcanism that could provide several outlets of natural gas. The Atherton tablelands contain outstanding assorted volcanic features, like Lake Eachan, which is a water-filled volcanic crater. Hence Northern Queensland should be endowed with many vortex fireballs events.

Tasmania is frequented with many UFOs. This is most likely related to possible gas emission from the Tasman Sea "line of fire", which is related to the Pacific Ocean "ring of fire", along which one should also see a high frequency of UFOs in the form of vortex fireballs.

Bob Young, a former manager of the electrical technicians at the Electrical Engineering Department at Canterbury University in around 1989-90 told me that Dr Ian Milner, the then Planning Manager of Electro

Corp (Christchurch, New Zealand) saw min mins in the desert while he was working in the outback. (Coleman, 1988a). The lights seemed to have a predilection for chasing or being attracted to cars. He described the min mins appearance as being like a tilley lamp in the distance.

What are min mins? I am of the opinion that the vortex fireball theory could well apply to these objects. The lantern appearance and distinctive movement leads me to this view. The vortex burner theory predicts that the localities where they would be found should coincide with sources of a flammable fuel, such as natural gas. The exploration of these sites with a view to discovering natural gas would test at least one prediction of the theory-that there should usually be a source of gaseous fuel to feed into the vortex.

The will o' the wisp.

The will o' the wisp is a type of UFO which was reportedly known by the ancients as "ignis fatuus" which means foolish fire (Clarke, 1988). It has been described as a flame-like light hovering over the marshlands on summer evenings. A popular view of the will o' the wisp is of a delicate light possessing a low energy. I have found at least one will o' the wisp account that appears to contradict this common perception. A young girl from the Vale Parish of Guernsey, in the Channel Islands, saw a robust "ball of fire" bouncing at high speed along the top of a hedge while she was cycling at night. This description is typical of the more energetic ball lightning events that I have encountered in Dr Singer's book and elsewhere.

Some scientists have explained the will o' the wisp phenomenon as mere pockets of burning methane or "phosphoretted hydrogen". Newton was of this opinion. However, such theories do not adequately explain the hovering maneuvers as they rise above hedges, or move vertically into air, or other complex motions. Will o' the wisps can move against the prevailing wind, and they can combine and recombine. They also exhibit a motion that seems to display signs of intelligence, to some observers.

The reputation for will o' the wisps possessing intelligence is based on observations that include the situation where the will o' the wisp is said to recede from an observer who approaches it, and to follow when he or she moves away from it. Therefore an explanation based solely on combustion is not satisfactory. A theory needs to explain why this light moves the way it does. The vortex burner hypothesis accounts for such motion because the motion of the observer disturbs the air currents feeding into the vortex fireball. It is well known in experimental studies of vortex-breakdown that the breakdown is super-sensitive to small air flow disturbances. Even small increases in air flow rate into the vortex will tend to move the breakdown downwards. Decreases in flow would send the vortex upwards. Therefore, if someone were to move towards or away from a fireball hovering nearby, the movement will cause a draught that will alter the position of the supposed vortex-breakdown.

Foo fighters

"Foo balls" or "foo fighters" were thought to be phantom flying objects that flew near fighter pilots in the Second World War. Hough and Randles (1994) in their book "*The Complete Book of UFOs*" speculated that the name "foo fighter" originated from a UFO sighting on the 23 November 1944 by an American night fighter squadron flying over France. The word "foo" is thought by some to be a corruption of the French word for fire, "le feu".

One theory of foo fighters was thought to be a secret menacing weapon of war. In a 1945 press release, it was speculated that the Nazis had fireballs and might be a form of ball lightning. Ball lightning was assumed by the press at that time to be solely an electrical phenomenon. The allies thought the Nazis had the scientific knowledge to be able to create ball lightning, while the Japanese were convinced that foo fighters were developed by the U.S.

Amazing aerial maneuvers of foo fighters were reported in and around the wing tips of airplanes. One feature of foo fighters was their apparent ability to travel alongside the aircraft at about the same speed the aircraft traveled at. It was not surprising that the question arose of whether the object had intelligence because the UFO was seemingly able to adjust to changes in the aircraft's speed.

The solution to the foo fighter mystery is to consider the idea that they were not independently powered. Instead it is more realistic to see whether they were directly connected with the aircraft's motion. I think the best explanation is that trailing wing or wing tip vortices were generated as the plane flew through the air, burning a gaseous fuel. There are obviously two wing tip vortices that are generated from the high pressure below the wing tip and low above. Air rushes from the high pressure below the wing tip to curl around the curved wing. The resulting air possesses angular momentum and a vortex is generated. Rotation is therefore clockwise/anticlockwise for the left/right wing tip vortex when seen from behind the aircraft. Turbulence does tend to break up the breakdown zone and lateral currents may also shift the position of the fireball. The breakdown burning zone is then able to move up and along the core. So a maximum of two wing tip fireballs on either side of the wing or trailing behind are predicted. I have seen a well publicized photograph of these foo fireballs in the correct position trailing from one fighter plane in World War Two.

Thomas Gold and others believe that methane or hydrogen plumes, being lighter than air, can be ejected up to heights as great as 50 000 ft with a lateral extent of 200 miles. This would be more than enough for the combustion effect to take place. That is, the foo fighters were vortex fireballs. If this solution is correct, then, by deduction the warplanes must have encountered some fuel gas cloud with spark ignition of the gas/air mix coming from St Elmo's fire, lightning or electro-static spark discharging from the metallic wing tips.

The fact that foo fighters have been seen in conjunction with aircraft appearing on wing tips, especially during World War 2, provides a real problem for ball lightning hypotheses based on metallic vapors. Take the following case, by Ingle (1971),

"In the second case, a bright ball appeared on the top surface of the wing outside the aircraft, made rapid movements to and fro for an appreciable length of time and then disappeared. It seemed quite unaffected by the air passing through it at 250 miles per hour, which does not accord with the theory that balls may be composed of vaporized metal."

What Ingle implied, of course, is that there will be a very sudden cooling effect from the very high air speeds flowing through and around the ball. This property probably rules out vaporized metal theories of ball in this situation. The combustion of a fuel gas in the vortex-breakdown regions of vortices shed from the surface of the plane's wings and or wing tips is the solution to the problem of foo fighters.

A second problem with metallic vapor theories is that they postulate an independent ball, not in any way self-propelled. How can motion with the aircraft be accounted for within such a theory? The moving air would merely blow any plasma ball away from the aircraft. The same problem is also faced by the fractal ball lightning explanation proposed by Smirnov (1987). Nevertheless, there is perhaps a very remote possibility that such an independent ball could be attracted or repelled electrostatically or attracted magnetically to an aircraft body and so resist the great air speeds flowing over the plane body. This is an unlikely mechanism to hold the ball in place amidst the high-speed air flows over the plane's fuselage.

Any proposed explanation in terms of a St Elmo's fire theory has to be able to account for fireballs that have the capability to weave in and out about the aircraft in all sorts of paths which St Elmo's fire is incapable of doing. St Elmo's fire does not move around independently but remains attached to the metal body of a plane.

A better solution to explain this foo fighter phenomenon is to envisage the fireball as part of the overall air flow over the plane, or perhaps the plane encounters a vortex which becomes latched on to the planes air stream and sustained by it. Vortices shed from the wings or wing tips, or fuselage, are set alight by an ignition source, such as a spark. Such plane-generated vortices do exist and have been the subject of much study. In fact the phenomenon of vortex-breakdown, as I have pointed out earlier, originated from the study of air flows over experimental aircraft wings.

Yakima Mystery lights

Greg Long in his book "The Yakima UFO Microcosm" described strange lights in the Yakima Indian Reserve. Bill Vogel, who was a fire officer from Toppenish, recorded many observations of the lights. Toppenish is around 125 kilometers in a direct line from Mt Rainier and 145 kilometers from Mt St Helens. Both mountains are major volcanoes on the Cascade Range of the north western coastline of the United States. It is also interest to note that the modern UFO phenomenon actually began with the sightings of nine disc objects by the pilot Kenneth Arnold in 1947 near Mt Rainier and roughly in the same area. The location of both sightings is fairly close to a major plate boundary and is well known for volcanic activity. This sighting could be a complete coincidence or else it could be connected to tectonic activity. To test the vortex theory in one aspect would be to look for a likely source of outgassing in the Yakima area probably along a minor fault line.

The Yakima lights were described as either a luminous orange-red sphere with a yellow-colored central region or white lights surrounded by a few multi-colored lights. It occurred to me that the colors of the first type of light with yellow in the middle and red on the outer zone suggests that the globe is not so optically thick in the visible spectrum because the center of the sphere can be seen. The other point being that the yellow region could be from luminous radiation emission from small carbon particles at a black body temperature of around 1600 degrees Celsius with a maximum wavelength corresponding to yellow. The red-orange color would come from carbon particles (perhaps soot) at lower temperatures of around 800-1400 degrees for a black body with a maximum wavelength corresponding to a red-orange color. The reason for the cooler temperature of these carbon particles may be related to the limited amount of air able to diffuse into the ball.

Rocket-shaped lights and columns of flames have been seen which apparently that did not leave scorch marks on the ground. To scorch the ground requires a minimum radiation level of kilowatts per square meters i.e. kWm^{-2}. For instance to scorch or burn trees the value would need to be around greater than about 10-15 kWm^{-2}. Certain parameters like the surface temperature of the flame, its emissivity and its distance to the ground as well as the view factor need to be established to calculate whether soil scorching would be possible. Just because there is no observed scorching does not mean that the object is not transferring heat by thermal radiation.

Strange sounds like turbines have been heard. This observation is consistent with a rotating turbulent air mass of a vortex combusting a fuel that could be natural gas.

The Marfa and Missouri lights.

Two more sets of mystery lights from the US are the Marfa and Missouri lights. The Marfa lights are observed in southwestern Texas near the Mexican border. Texas is rich in natural gas and could provide a natural gas source of fuel gas for the Marfa lights. These lights have been seen for well over one hundred years and have been described as white lights that can pulsate, disappear and reappear and they can be seen in groups bobbing around. Despite numerous attempts, a convincing explanation has yet to be widely accepted. I believe these lights could be explained by the vortex fireball hypothesis.

The Missouri lights are seen along the Missouri-Oklahoma border at Hornet, near Joplin. The lights do all the usual things that a mystery light can do. They have many colors, they split and dance around. Some are transparent and objects like trees can be seen through them (Wilson, 1997). One spinning orange ball in the 1950s went over a car and continued down the road. Of interest are the flashes of light seen in the 1811 New Madrid earthquake, also in the state of Missouri, which is most likely from natural gas combustion. The Missouri lights show all the signs of being vortex-breakdown fireballs.

USOs-Unidentified submerged objects.

USOs are unidentified submerged objects which some authors have defined as fiery aerial objects that come down and hit the water sending up a column of water. Basically all the physical characteristics of USOs can be attributable to vortices (i.e. water spouts). For instance, vortices are well known to raise water as they touch down on the water's surface. A USO report in Randles (1982) described unexplained red lights above the sea in Shag Harbor in Nova Scotia on the 4th October 1967. These lights flickered on and off and then began to decay in light intensity. These lights then combined to form a white light that moved up and down in the sea. A later examination revealed a small area of bubbling water and some yellow-colored foam.

The bubbling water described in the above quotation is most likely natural gas bubbles. I suspect that the gas bubbles at the time of the UFO would have been natural gas ascending through the water from some source like a fissure in the sea floor. The other possibility is a biological decomposition process that is much less likely given that such "marsh gas" usually occurs in stagnant bodies of water. Natural gas is probably the culprit and source of fuel gas for this USO. In my opinion, USOs are just another name for a type of UFO seen at an air-water interface. USOs are here identified as burning waterspouts.

EXPLAINING UFO SHAPES AND THEIR CHANGES

UFO shape changes can be dramatic. The filming of the Kaikoura UFO demonstrated that rapid changes from a bell shape to the classic flying saucer shape can take place over a fraction of a second. The vortex hypothesis can account for these changes in the same way I explained ball lightning structural changes. The basic idea derived from the hypothesis, is that shape changes are alternations in the flame boundary brought about by changes in the air flow into vortex-breakdown. This in turn brings about a change in the air velocity distribution within the vortex such that certain regions of the vortex will combust while others parts will be non-combusting, as I mentioned in the case of ball lightning.

If there is going to be any combustion at all within vortex-breakdown it will be most likely located around the stagnation region where the air speeds are substantially reduced.

Vortex-breakdown assumes a variety of shapes. The speed of the air flow provides an upper limit on whether combustion would take place in certain local regions of the vortex-breakdown. This situation means that isolated regions of combustion can take place instead of the usual spherical form. Hence, such shapes as tori, ovals, saucer, inverted ice cream shapes, and even more complex burning regions, could be envisaged using this idea.

Changes in geometry from a spherical or oval form to a disc shape could be to do with localized combustion in the region of the stagnation region just above the base where there is lateral expansion of the vortex core. Because of the varying air speeds into atmospheric vortices, rapid shape transformations like that seen with the Kaikoura UFOs are then possible.

Tubes and cylinders

The book "*UFOs and how to see them*" written by Jenny Randles, in 1992 contained a few references that I think can be explained by the vortex burner theory. Randles (1992) cited a UFO seen at a place called Tomakomai, close to Sapporo, the capital of Hokkaido, Japan. A man saw an orange-colored light in July, 1973 (in summer) descend in a spiral path down and remained a meter or so above the sea's surface in the local bay. Suddenly a tube extended from the object to the sea and drew water up into the UFO like a vacuum cleaner through a sucking process.

The evidence that this particular sighting and possibly many more like it are vortex fuel burners is based on the following evidence. The description of a "tube" certainly seems to suggest a vortex funnel sucking up water. UFO observations of glowing cylinders and glowing spheres with cylinders emerging from the bottom are consistent with the primary vortex funnel leading up to the vortex-breakdown recirculation region. Tomakomai is in Hokkaido and could have been a site for natural gas. Gold (1987) said that Hokkaido is the southernmost part of an arc which reaches up to the Kurile Island where there is vigorous outgassing taking place. On the island of Hokkaido itself the local coal mines are apparently little mined because of the presence of natural gas diffusing through the coal seams providing a potential source for hazardous explosions.

Funnel and sphere connected.

In the book *"The UFO Experience"* by the scientist Dr J.A. Hynek, there is an exceptional photograph of a UFO shaped like a globe. The image comes originally from a film by the television newsman Bob Campbell taken around 3 am on the 2nd of August 1965, at Sherman, Texas. An image of the photograph is available at the following website http://www.temporaldoorway.com/ufo/analysis/campbellphoto. The reason why it is so exceptional is that there are features which correspond to what would be expected if the object was a vortex burner. My interpretation is that there is a slightly curving luminous funnel region leading into a bright spherical zone. The glowing regions being located in the vortex funnel leading into vortex-breakdown is unexpected since one would expect no flame would survive here because of the high speeds which would blow out any flame. If there is to be a combustion flame then low air speeds are required. Thus these lights may be localized turbulent eddies with recirculation zones with much reduced air speeds (1ms-^{1})-sufficiently low for combustion. If this were the case, then such a situation would clearly be scientifically important, since such an observation may not have been reported of an atmospheric vortex.

The original photograph of the globe of light had a halo effect and raises the distinct possibility that the visible boundary may be a result of a halo effect from a bright light source. This suggests that light emitted radially from a bright source inside can overwhelm the original flame envelope. This is analogous to looking at an ordinary light bulb where an observer can see the bright, round glow but not the filament itself. This halo effect could take place in some ball lightning episodes and may help to explain why some balls are spherical. This explanation maybe a way of accounting for a spherical glow other than by a spherical flame produced in a swirling vortex.

A luminous vortex over the Canaries

One of the best documented sequences of rapid UFO shape changes is contained in the book *"The Age of the UFOs"* edited by Peter Brookesmith. The book contains a temporal sequence of photographs taken in March 1979, by Antonio Gonzales Llopis. The pictures display a huge luminous UFO and different stages hovering over the sea in the Canary Islands area south west of Spain. Its peculiar shape is a distinctive departure from the traditional disc or globe. The UFO showed a remarkable series of transformations. One shape of the UFO is rather like a balloon. In this case the "balloon" is fully "inflated" but with the throat of the balloon (the lead-in vortex core) dangling down. The main stagnation zone would be located just above the vortex core inside the breakdown zone (the "balloon"). Another photograph showed the UFO with the round shape below converges to a sharp apex like that at the top of a cone. I puzzled over how the vortex-breakdown hypothesis could explain this sharp cone. Do any vortex-breakdown experimental forms match this? How is this cone to be explained? On the 3rd February, 1998, I found that another conical form of vortex-breakdown had been reported in the experiments of Sarpakaya (1995) and also by Khoo et al (1995). This conical vortex-breakdown may explain the case of the teardrop shape of the Canary Island UFO.

My interpretation of the orientation of the Canaries Island vortex is that it appears as though the large teardrop shape is nearer to the observer than the vortex core below it. The vortex is thus sloping away from the observer. The original photographs show a golden color that I put down to the sodium emission from sodium ions present in the sea air entering the vortex. This would be expected since the UFO was seen hovering over the sea.

The UFO that appeared over the Canaries is the best example I have seen where the main form (the balloon form) is so similar to the shape of vortex-breakdown I have seen in my experimental studies of vortex-breakdown.

Orbiting fireballs around a "mother ship"

Some UFOs have several fireball objects circling around a common center. The flaming orbs show a surprising ability to coalesce into a single ball, paralleling published ball lightning observations also split and recombine. This striking similarity of ball lightning and a type of UFO reported can be explained successfully with the vortex burner hypothesis by appealing to the "vortex splitting" behavior I mentioned earlier.

Consider the "mother ship" UFO behavior cited in a letter written by a woman to well-known UFO investigator, Dr Allen Hynek. The woman was from a southeastern state and did not wish to be named. On 2 February 1966, a UFO was seen hovering at night time above trees in the yard of a neighbouring house. Small explosions were seen and heard coming from the object. The object was described as silver-colored diamond shaped UFO with about twelve or so smaller red, green and blue-colored spheres were flashing colors like some stars do in the night sky, while orbiting about the central object. Although it was generally in one place, the object rocked to and fro. Under the vortex theory the balls probably rotated in a near vertical axis, although this orientation was not specifically stated. The spheres were generally orbiting about the main UFO but appeared to have their own trajectories. This aerial display stayed in the same location for around three to four minutes when it unexpectedly moved away from the observer to the northeast (Emmeneger, 1974).

Two observations in the above sighting can be explained with the vortex burner hypothesis. The smaller balls rotating in an orbit about a common diamond-shaped object is a prime example of a multiple vortex system, which is a scientifically documented behavior of vortices, known as "vortex splitting, as I discussed earlier. The phenomenon has been seen in the Ward laboratory vortex generators, where smaller subsidiary vortices rotate about a common center. In the above UFO observation there was a central vortex about which the other daughter vortices rotated. The second observation is also noteworthy. The witness drew a sketch of the path of the object which showed that it moved away in a simple arc path, which is, indeed, typical of the motion of real vortex splitting systems in nature (Gibilisco, 1984).

Gas lamps and lanterns

UFO sightings are common in New Zealand. In the book *"Aliens over Antipodes"*, by Murray Stott (Stott,1984) furnished a report by Fr Peter Durning, a Dominican priest, who recollected his experiences as a pilot on an NAC flight from Wellington to Christchurch in the early 1950s.

As he was getting close to Kaikoura he saw an orb of light slowly moving downwards from a cloudless sky. At first, he thought the object was a weather balloon but the light diminished like the flame from a gas lamp mantle. The portions of light "poured" off from the main body of light and the whole thing vanished out of sight. No signs of the object were left.

This vivid observation would seem to indicate a chemical combustion process mainly because of the way the light behaved like a gas lamp flame which slowly decayed and eventually extinguished leaving no trace. In my opinion, this UFO light really was a gas lamp in the sky with function of the lamp body and wick replaced by a vortex. Therefore, this observation, in my opinion, is consistent with the vortex burner hypothesis.

The lantern or tilley description would appear to be a common occurrence. UFOs seen in Papua New Guinea and min mins in Australia have all been variously described as lanterns, lamps or tilleys. The jack o' lantern is also based on this theme and is another name for the will o' the wisp that is an unexplained light.

When I visited Japan I was struck by the physical design of these Japanese lanterns. On a closer examination they had a geometry which reminded me of ball lightning and UFO forms. In some of the lanterns the top cylinder of the lantern "globe" is wider than the bottom cylinder. It would not surprise me if perhaps the Japanese lantern seen hanging in some older streets may have its origin in these vortex fireballs. Japan is a located at the intersection of three major plates with a plentiful supply of natural gas. In countries with natural gas these spherical vortex flames may have provided inspiration for myths and legends of the culture. In that case the Japanese lanterns maybe festive representations of vortex fireballs.

COLOR OF UFOS

How is the color of a UFO to be explained within the context of my hypothesis? I discussed ball lightning color in Part I. The same reasoning would obviously apply to UFOs that are vortex burners. One would expect to find that the color of the UFO would depend on what wavelengths of light are emitted from the combustion flame. The color could be quantum light from transitions of chemical species in the flame, bearing in mind that flames radiate neither as grey nor black bodies. The other possibility is that trace impurities will produce the dominant color. For instance, a pre-mixed stoichiometric flame at atmospheric pressure in air would usually burn with a blue color without impurities. If salt (largely sodium chloride) is added to the flame the color would change to orange-yellow. Such an effect is easily demonstrated with a simple Bunsen flame test. Strontium salts give a crimson color, whereas sodium would yield a yellow color, and copper a green color and so on. If the flame was non-stoichiometric, the flame would be yellow from light emission from carbon particles.

Green fireballs in the desert

Another piece of indirect evidence that UFOs have a distinctive color because of a metallic species present in particles in the atmosphere was cited in the book "*The Complete Book of UFOs*". In Chapter 8 on pages 86 to 92, there is a discourse on the subject of green fireballs. Apparently during November and December of 1948 there was a spate of sightings involving green fireballs near Los Almos. The people who observed these fireballs were trained observers and included air force pilots and scientists.

The sampling of airborne of impurities in the region through which the UFO flies would provide stronger evidence, though not conclusive, for light emission from these substances present in a combusting flame. UFOs seen over the sea would appear yellow because of the presence of sodium in sea aerosols. The Kaikoura UFOs, for instance, were an orange-yellow color consistent with combustion of a fuel gas in a medium likely to contain sea salt, rich in sodium ions.

Consider the investigation of green fireball UFOs by a world authority on meteors, Dr Lincoln La Paz, of the University of Albuquerque who failed to locate debris material from the suspected meteor crash site. He found that the UFO moved at much lower speeds than meteors, and there was no sonic boom. The meteor did not leave a spark trail at the rear. La Paz's conclusion was that the fireballs were not meteors and he suggested that what was seen was not any known natural phenomenon.

A significant piece of deduction work, which provided a clue to the puzzle, was La Paz's conclusion that the green color of the fireballs was due to a high copper content. The evidence to back his claim up was that copper dust was found on the ground beneath the flight paths of the green fireballs. This piece of evidence, though not conclusive, ties in very well with the vortex-breakdown combustion model of ball lightning and is consistent with it. This does not rule out alternative flame combustion models that might account for the green color, or that the copper just happens to be in the atmosphere at the time.

WHERE ARE UFOS TO BE FOUND ?

The vortex burner hypothesis predicts that many of the unexplained UFOs that people have observed in the past should be generally found in the same kind of locality as ball lightning. These UFO sightings should coincide with areas rich in combustibles, like natural gas or hydrogen sulfide, and, to a very much lesser extent, hydrogen gas. Such areas would be fissures along the world's fault lines, as well swampy regions where there is an abundance of flammable gas. As I have said earlier, some UFO researchers, such as Persinger, have claimed a trend where UFOs are more frequent along seismic zones, such as earthquake fault lines. This connection is fully accounted for within the theory provided the gas is flammable. But such correlations and UFO theories directly trying to explain this correlation have not gone far enough. More specifically, increased UFO sightings of the fireball kind should coincide in areas of high vortex production intersecting with active gas emission areas and with a source of ignition. Swampy regions, for instance, could therefore be included. Both these areas would provide favorable conditions for ball lightning creation.

Canterbury, New Zealand UFOs

Consider UFOs seen in and around the Canterbury area. If UFOs can hover over this terrain there must be a source of fuel to power these vortex fireballs. At first I thought Canterbury would not be well endowed with natural gas seepage, but an article in the Christchurch Press, reported by Watson (1997) suggested otherwise. The director John Holland of Indo-Pacific Energy, an international oil prospecting company said that the kinds of oil and gas bearing rocks that are observed in Taranaki just happens to exist in the Canterbury Basin. Various gas leaks have been detected in this area. Such gas seepage in the Canterbury basin could be important in providing fuel gas to feed into UFOs seen at various times hovering over the Canterbury region.

Spinning light on the San Andreas Fault

From the point of view of the vortex burner theory, a positive UFO-fault line correlation is to be expected with the vortex fireball hypothesis. More UFOs should be observed in areas where there are active earthquake faults that emit natural gas. Barring the lack of hard data on gas detection in some UFO cases, the association of a UFO with this specific fault does point in the right direction. Nevertheless, the confirmation of natural gas detection at this UFO site would support my hypothesis.

Devereaux illustrated his UFO fault line correlation proposal with a UFO object that was seen in 1973 along the Pinnacle fault line in California, near the famed San Andreas Fault. The observer was a physicist called Dr David Kubrin. The vivid description is typical of what I would expect if someone was to see a spinning vortex fireball close up. What Kubrin actually saw and photographed was a type of light on the Pinnacles fault line a few miles from the well-known San Andreas Fault. The light was able to produce what Kirin described as "shock waves" around the object. The eyewitness observation that the light was rotating on an axis also supports the vortex combustion hypothesis.

The "shock waves" ahead of the object and the streaming down around it, I would interpret as the movement of air in the vortex, possibly enhanced by the heat from the burning vortex, in the same way that hot air rises from a hot road in summer. The rotation described in the sighting is certainly indicative of a vortex. The light diminishing and then extinguishing is consistent with a burner running out of fuel gas. The continuous reduction in light to its final quenching is a feature consistent with a combustion-based hypothesis, as is the observation that it took place in a seismically-active location along a fault zone. This is the first time I have heard of a spinning earthquake light seen at close quarters.

Levelland and White Sands UFOs-why there?

The vortex burner breakdown hypothesis predicts that a greater frequency of UFOs should be seen in areas that coincide with natural gas sources, such as natural gas fields, marshlands and areas where vortices are frequent. There is available information on the location of natural gas reservoirs in nature to test this correlation. I decided to check this idea out with a specific UFO case to see if I could match the location of well known cases of UFO events with a known gas field located in the general area. I did find a match. Take a well known sighting of UFO in the form of a ball of fire which stretched across the width of a road, near Levelland Texas (Randles, 1987). I proceeded to find out if there were any natural gas fields in the vicinity. I consulted Gold's book and found a detailed description of the so-called "Hugoton-Panhandle" fields of Kansas and Texas. The Panhandle field extends into Texas straight towards Levelland. The fields are rich and with a high proportion of methane (Gold, 1987. Here was clear evidence of the existence of gas fields in the area. I discovered that White Sands is only 200 hundred miles west of Levelland, which also happens to be a very well-known UFO site for which vortex fireballs could be supplied with natural gas from the extensive gas fields.

The demise of the will o'the wisp

The story of the will o' the wisp, or "foolish fire", and its decline in certain regions in England, such as Dartmoor, is interesting and relevant to the link between UFOs and marsh gas abundance. The will o' the wisp is less common in England than it used to be (Clarke, 1988). The will o' the wisp was usually seen over marshy ground, church burial grounds, lakes and lochs which would have been likely locations for the occurrence of methane.

An example of a will o' the wisp sighting which took place in Northern Ireland is instructive and demonstrates yet again, motion against the wind. In the year 1912 the Earl of Erne from Crom Castle, County Fermanagh, Northern Island, in the London Daily Mail described the appearance of a strange light over Loch Erne. The light had been seen for six or more years by many observers. The light was yellow like that from a car light. The light illuminated objects within a certain radius. It had the capability to sudden appear and then just as easily disappear. The light could move along with or against the wind while remaining above the lake's surface.

Why has there has been a significant reduction in will o' the wisp sightings in marsh areas? The simple answer is that the marshes have been drained of water depriving the fireball of a source of marsh gas. Anaerobic biochemical decay of vegetation in watery masses, such as lakes, is what produces the marsh gas. Bacteria on sulfates are known to produce hydrogen sulfate. The effective removal of the biological decay conditions required to generate fuel gas effectively eliminates the necessary chemical energy for the vortex fireball. Of course, the non-combusting will o' the wisp is then merely a vortex free to continue wandering through these country lands. It will be then be a largely invisible phantom unlikely to attract attention.

The Bermuda Triangle UFOs

The well known mystery of the Bermuda triangle in the Sargasso Sea, off the coast of North America is linked with the disappearance of aircraft and ships. No one theory has satisfactorily explained the sightings of unexplained UFOs and fireball objects with tubes extending to sea level. Such accounts also might be explained using the vortex burner theory. The possibility is that natural gas emission in large plumes into the atmosphere over the Sargasso Sea area from areas around the Mid-Atlantic ridge. It is known that along the eastern boundary there is active volcanic activity in the form of seafloor spreading.

EFFECT OF Ufos ON VEHICLES.

Some UFO reports show that vehicles can be affected in various ways when they encounter a UFO. Jenny Randles in her book *"The UFO Conspiracy"* described a well known encounter of a fireball "sitting" on a road near Levelland, Texas, just after midnight on 3 November in 1957 (Randles, 1987). The UFO was seen at 12.15 am and apparently it stopped the car. The yellow UFO seen by Mr Frank Williams of Levelland performed a most unusual thing. As the ball pulsed like a strobe so too did the headlights of the car. The UFO then suddenly took off with a roaring noise and the car returned to normal.

There are a number of different types of possible UFO effects on the engine and lights of vehicle that have been reported. I will attempt to explain these effects using the vortex fireball explanation. I am aware that the reports may be incorrectly recorded, though probably unintentionally. There have been a sufficient number of such cases to warrant an explanation. I will briefly explore the ability of the vortex burner to explain such events.

In some encounters both the engine and the head lamps fail while in others, only the engine fails, and the lights remain on. I have even come across a UFO event in which both the engine and lights fade but not totally. Then are events where there are no noticeable effects at all such like the case of a close encounter of a car and a fireball UFO whose diameter was the width of a road. Apparently, if the observations were authentic, any theory would need to be capable of explaining all these variations. The vortex fireball hypothesis could explain these vehicle effects either through some mechanism involving electro-magnetic fields generated by the vortex using charged particles inside the fireball, or perhaps, a more likely mechanism associated with the air flow of the vortex.

Let me briefly explore the first possibility. Electrostatic or electro-magnetic fields emanating from charged particles in the vortex-breakdown region might conceivably interfere with the cars electrical system-perhaps shorting across key terminals like that mounted on the ignition coil. However there is large problem to overcome. If there was electrical field generation by the vortex, I would expect there to be little effect on the car, even if a strong field was present. This is because most vehicles have a metallic hood over the engine, which would effectively act as a Faraday shield preventing any electrical interference to the inside of the vehicle's engine. The observation of headlights going out does seem to suggest electrical shorting out of some kind-perhaps by charged or conducting particulate matter swirling around the inside of the vortex and into or under the car's bonnet. I would consider this scenario much less likely.

On the other hand, the low pressure of the vortex could affect air flow into the carburetor that may account for some cases where the engine stalls. If the lights are dependent directly on a generator and not the battery then this provides a reason why the lights would go out. However, most cars have a battery and even if the car's engine stopped, there is always the battery to keep the lights going. Obviously each make of car and it's associated electrical and fuel system needs to be taken into account in the light of the hypothesis to come to better idea of what actually happened.

It is interesting to note that when these fireball UFOs "fly" away the effect also vanishes which is consistent with some temporary field effect, be it the air flow effect or an electro-magnetic effect. Intermittent stalling may make the headlights fluctuate in intensity in synchronization with the pressure changes in the vortex fireball and emitted light from the UFO. The exact mechanism needs further elucidation.

Then there are even cases where no effects whatsoever are reported on the headlights despite the car actually being physically lifted off the ground. I am thinking here of the Knowles family car involved in the Nullabor plains UFO incident. Is it that modern day cars have better electrical shielding and fuel/air systems more resistant to large air flow perturbations than that of older cars?

The Hessdalen lights

Smirnov (1994) a well-known ball lightning researcher, wrote a report on an international conference held in Hessdalen, Norway in March 1994. The conference examined "Long-lived phenomena in the atmosphere". Unexplained lights have been seen in a 12 km long valley in Hessdalen in Mid-Norway close to the Swedish border. From 1981 the inhabitants of this area reported seeing these lights. Smirnov stated that the lights were three basic types. The first type was a yellow ball that exists for 1 to 2 hours and every 5 to 10 minutes it changes location. The second type of light was a bluish-white color that can flash and is seen high above the valley and surrounding mountains. A third type was a formation of lights connected and moving together. Smirnov did not say what the geometric arrangement of such lights was, but Randles (1982) has a photograph of such lights on page 99 that showed 5 prominent lights appearing to connect together in a regular geometric pattern. The lights were seen hundreds of times between 1981 and 1984. In 1983 a scientific project was established to study the lights. The equipment used was high tech. It included video cameras, infra red viewers, spectrum analyzers, radar, lasers and Geiger Muller tubes (used to detect radioactivity).

Ashpole (1995) wrote *"The UFO phenomena-A scientific look at the evidence for extraterrestrial contacts"* and cited the Hessdalen scientific project. His report on the findings was interesting. Apparently these Norwegian UFOs had a preference for taking place during mid-winter when the mountains and valleys were covered in deep snow. In winter the climatic conditions are more conducive to the build up of natural gas clouds close to the ground. The weather may be more settled so the air is still enough so there is little or no dissipation of these clouds. At the lower temperatures, the density of natural gas is increased which means it will be less buoyant than at higher temperatures.

Devereaux (1988) reported that the Hessdalen lights appear in various sizes and colors. A typical shape of the light is that of an upside down pine tree. These lights are able to move along mountain ridges. The luminosity is variable and they can become bright and then decrease in brightness to become optically thin or transparent. The upside pine tree UFO has two main parts. The trunk and the triangular region represented trees foliage. The triangular "foliage" region is the vortex-breakdown while the upper "trunk" is the vortex funnel.

The characteristics of the UFO near Hessdalen led me to suspect that eyewitnesses are actually reporting vortex burner events. The predilection for the Hessdalen lights moving along a mountain ridge is characteristic of atmospheric vortices which have also been seen to move along ridge crests. Norway does have abundant natural gas fields that under certain conditions could fuel vortices in the Hessdalen valley. The observation that the lights are seen in winter on snowy surfaces could be related to smooth snow or ice surfaces that would be ideal for vortex creation. Devereaux (1988) stated that faulting does exist in Hessdalen and that there is an increased frequency of the lights with increases in crustal movement. This would be another observation in favor of the vortex burner theory.

The Hessdalen lights have two more properties explained by the vortex fireball theory. When a laser was shone on some UFOs it seemed to respond by changing its light intensity or by moving. The only way this seems possible is that the laser beam heats a local region of the vortex-breakdown and alters the air flow. It is well known that vortex breakdown is very sensitive to small changes in an air flow. A small heating effect could shift the vortex breakdown along the central axis of the vortex. The second observation is the shape of the Hessdalen light. After examining one UFO photographic image on a Hessdalen web site, I found the form was essentially like that of the Kaikoura UFO, and incidentally like a widely acknowledged photograph of ball lightning by Norinder cited by Singer (1971). The flattened top (in this case also with a narrow tube), rounded sides coming to a rounded apex at the bottom (see also the drawings of the Kaikoura UFO and Norinder ball lightning).

The 1978 Kaikoura UFOs

Now UFOs have been seen in every part of the globe. Even in New Zealand where I live, UFO reports periodically surface in the media. George Graham, a resident of Mt Pleasant in Christchurch saw an oval ball coming from the east. It emanated a bright bluish-white light, which he estimated to be the size of a soccer ball. This object changed direction and proceeded in a southerly direction (Keanan, 1990). Mr Graham said: *"I'm no UFO buff, but when you see something like this, there has to be an answer."* Several other Christchurch witnesses also saw the object. A representative from the Canterbury Astronomical Society speculated that it was a meteorite, but Mr Graham countered, *"I have seen meteorites before and it did not look like one to me."* Other learned commentators on meteors have said the same thing. Dr Gerald Kuiper, then Director of the Yerke's Observatory in Williams Bay, Wisconsin believed that some so-called "meteors" bore no similarity to conventional meteors. The description of these UFOs points to a completely different phenomenon and conflicts with the meteor hypothesis. In the original press release he stated that he saw a change in direction of the blue fireball which is something that meteors do not do. The UFO was first seen moving in an easterly direction as it came in from the sea. It then went across his field of view and shifted to the south. Meteors simply do not dramatically change direction in this way. They move in well-defined parabolic trajectories, except when they disintegrate into fragments. However meteor fragments do not suddenly change direction like the object in Mr Graham's account.

Strange moving lights have been seen around the Christchurch Port Hills. UFOs have been sighted in such locations as the Estuary, Godley Head, and Lake Ellesmere. In one local newspaper press story, a young boy saw a strange fireball at Cass Bay. There have been many other such UFOs seen, but most have gone largely unreported. The cases that make it into the media represent only the tip of the iceberg.

Three years before the famous 1978 Kaikoura UFO sightings, Hervey (1975) compiled a useful number of sightings from Australia and New Zealand in the book *UFOs over the Southern Hemisphere*. In Corrigan, Western Australia, Hervey reported on 203 sightings in 1969 alone. The year 1969 is also a significant year for New Zealand sightings and he described eighteen New Zealand UFO events in chronological order. Discs, cylinders and flaming spheres were seen. On 6 June a couple traveling in their car on Highway 50, near Gwavas Station in the Hawkes Bay saw a strange football-sized light in the sky. It was described as being like a ball of fire, with a green tail.

Hervey wrote about contemporary sightings in the twentieth century but such sightings are not new. There are earlier accounts of fireballs in New Zealand. I came across a *NZ Truth* (Oct. 10th, 1988) article of a fireball object moving down a bush-clad valley in the Ureweras. This unexplained object was seen and reported by a British colonel, St John, who fought the Maori chief, Te Kooti with members of the Arawa tribe. The Maori allies of St John thought it was a secret weapon of Te Kooti and that it would signal their defeat in battle. Another wartime interpretation of a UFO is found in the famous foo fighter phenomenon of the Second World War.

But it is one of the most publicized UFO cases that has really grabbed the limelight. This was the famous sighting and filming of unusual lights off the coast of Kaikoura New Zealand on New Year's eve 1978/1979. A detailed description of the famous Kaikoura UFOs is contained in the book "The *Kaikoura UFOs*" by Captain Bill Startup and journalist Neil Illingworth (Startup and Illingworth, 1980). What is remarkable about these sightings is that they were extensively recorded on film by trained TV personal. Dr Bruce Maccabee, an American physicist specializing in optics, subjected this film record to scientific scrutiny. He subsequently analyzed the film and wrote a scientific paper on the subject.

Several pertinent features of the case lend credibility to the sightings. First, they were recorded on film while independent radar groups in Wellington and Christchurch verified the UFO location. In addition, several trained and highly experienced observers saw the objects. Not only that, but the aircrew had been on this particular flight several times previously. These factors are important when due consideration is given to the credibility of the eyewitness accounts. As seasoned operators, they could quite clearly distinguish the difference between what they saw, and other more familiar objects that were invoked at the time.

Theories put forward for this Kaikoura UFO included: Venus, squid boat lights, train lights, car lights, and beacons. It is important to realize that the crew was already experienced in identifying these things. They were familiar with them. It was suggested at the time that the lights were merely inversion effects from squid boat lights or car lights originating from refraction atmospherics from northwest conditions. The refraction hypothesis lacked the ability to account for what the observers actually witnessed. Are these lights detected on radar? Do they flare up, shrink or throw out flames or perform unusual acrobatics? An official report, called the *Ireland Report,* by a New Zealand government research institute, called the DSIR, did not adequately explain these UFOs, according to the authors Captain Startup and Neil Illingworth. I think they are right.

What they saw on that night was something quite different than anything they had ever encountered. As the author of the book says, many of the unusual lights were seen out at sea which would rules out the car or train lights explanation. As far as I know no one has come up with a workable explanation that would successfully explain what the original observers saw on that night.

Now to recount the actual event that took place on New Year's Eve in 1978. A four-engine Argosy aircraft piloted by Bill Startup and Bob Guard was flying the usual route from Blenheim to Wellington and then back towards Christchurch along the Coast of the South Island. The first sighting took place a few minutes after midnight near Cape Campbell on the while the plane was flying south. More lights were seen between Cape Campbell and Kaikoura and were described as appearing and disappearing in an unusual fashion (page 96).

The lights moved up and down as well as laterally and were able to move alongside and travel with the Argosy aircraft that was reportedly moving at 215 knots. If the Kaikoura UFO was a vortex fireball then it was most likely a large independent vortex not generated by the air flow of the aircraft's motion. The bright UFO lights were capable of generating bright beams inclined at 45 degrees to the sea below. They were able to turn off and then start again (page 102). There is some similarity of this beam effect with what I have described in connection with the ball lightning object of Zou (1989) and the Andes effect light. The vortex burner theory can explain this beacon behavior by reflected light down the vortex column.

A most startling observation took place on the northbound return trip back to Blenheim as the Argosy aircraft was climbing in height just out of Christchurch. The UFO appeared as a large orb. It was filmed and a photograph appears on the cover of the Startup and Illingworth book. The film crew was better prepared this time with faster film and they were more familiar with operating their camera on the flight deck. An examination of the film of the "squashed orange" orb revealed that the shape and trajectory of the object could change dramatically between frames. The frame speed was 10 frames per second which means that the time interval for this change took place within 0.1 of a second. One of these shape transformations was to a bell-shaped form. There were disc-shaped objects as well, very much like the classical flying saucer, as well as drum shapes in the film frames.

The color of this large orb was described as orange in color by Bob Guard, who likened the light to the color of halogen lamps seen at Fairlie in the South Island. This description is consistent with the color plate of the object in the book that showed a yellow-orange object. The large fireball followed the aircraft was similar to descriptions of the World War Two foo fighters balls of light. They lost sight of the unexplained light but Later, near Cape Campbell still more lights were seen in a final display.

The Kaikoura lights had the classic UFO shapes such as the "drum" or "saucer disc" which can rapidly change. According to the hypothesis flame shapes regions within vortex-breakdown can change over small time intervals as a result of variable flow of air into the vortex-breakdown recirculation region.

The 1978 Kaikoura UFO- same basic shape as ball lightning.

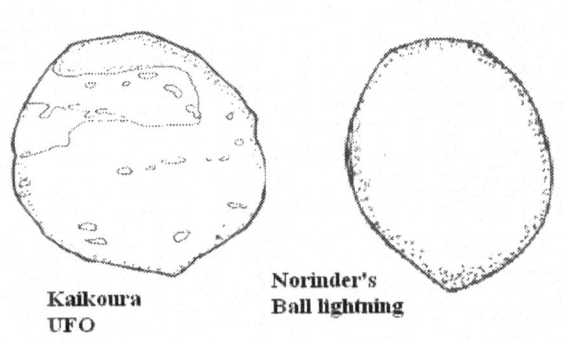

Kaikoura
UFO

Norinder's
Ball lightning

Figure 19: Note the similarity between the Kaikoura UFO shape and Norinder's ball lightning. Vortex breakdown is suspected as explaining both

When scrutinized the Kaikoura UFO photographs in the book by Startup and Illingworth I discovered that the shape of Kaikoura UFO matched the shape of the authentic photograph by Norinder. Incidentally the original photograph came from Schneidermann (Singer, 1971). I think there is a physical reason underlying this similarity. The shape in both cases was a basic ovoid yet flattened at the top and more pinched in form at the base like a pear shape. Actually the Berger photograph of ball lightning reproduced in the encyclopedia Britannica also has a flat top and apex at the bottom. Now if the reader will look at the Canaries Island UFO they will notice the thin wavy tube heading towards the sea surface at the base of the ball. This is a shape that is like a fingerprint of vortex-breakdown. As a general rule in vortex-breakdown research the diameter of the funnel leading into vortex-breakdown is one third of the diameter of the top.

The reason why the Kaikoura UFO and the Schneidermann or Norinder ball lightning shapes possess a common appearance is that the observer is seeing the region where the ball joins the "umbilical" cord of the vortex core. Ball lightning or UFOs with tails hanging down are common and are interpreted as a luminous vortex core leading up to the vortex-breakdown. The flattening at the top is probably the re-entry region of the vortex core downstream of the globular recirculation region which experimental studies have shown to be wider (about three times the diameter) than the section of vortex leading into the vortex.

Seismic activity and the Kaikoura UFO

I needed to find out where the fuel gas that supplied the Kaikoura UFO could conceivably have come from. I think the source of fuel gas could have arisen from a major release of natural gas before seismic activity along an active fault, such as the Clarence fault, or other active faults, either inland, or at a series of locations along the sea floor. It must have been a sizeable release because at least one of the UFOs was reported to have traveled a few hundred kilometers. Fluctuations in the supply of the fuel gas might explain the disappearance and reappearance of the UFO. However, some form of heat source must continue on a small scale to reignite other areas within the vortex-breakdown to flare up again.

To further support the presence of a natural gas cloud to fuel the Kaikoura UFO, I looked for evidence of major seismic activity along the faults, in the area around the time of the sighting. The fact that the UFO traveled parallel to a well known active fault line, pointed to such a possibility. I decided to check the New Zealand Seismological Observatory data to see whether there had been in seismic activity in the Christmas holiday period from December 1978 to January 1979. Although as seismologists say, seismic activity in the form of small tremors is going on all the time, what I was looking for were earthquakes of at least moderate magnitude. My assumption is that tectonic activity might have released natural gas along the fault line at the time and was signally by seismic tremors. As I have said, there is a precedent for this natural gas-earthquake association, which was reported in Gold's book and cited, in other scientific literature. Of course, not all earthquakes will emit natural gas. I suspect that natural gas was emitted during the 1978/79 Kaikoura UFO event.

When I read through the 1975 to 1978 *NZ Seismological Bulletin* a number of different earthquakes were recorded all over New Zealand. But the very last recording of that data period was on November the 30th and located at 21 hours 46 minutes and 38.7 seconds at a depth of 8 kilometers (Smith, 1979). An earthquake was recorded with an epicenter at latitude of 41 degrees 72 seconds, longitude 174.12 and with a magnitude 4.2. This tremor was located south of Blenheim and was felt as far away as Wellington. I plotted the epicenter on a map of New Zealand and I found it was located out at sea, about halfway between Cape Campbell and the Clarence River mouth.

I then studied the bulletin from 1979 to 1982 period (Smith, 1983) and found that at the start of this record, an earthquake in Cook Strait took place on the 13th of January, with an epicenter at 06 hours 14 minutes and 35.1 seconds. The depth of the earthquake was recorded at 33 kilometers and had a magnitude 4.9. The records showed that only two earthquakes took place from November 30, 1978, to the 13th January 1979. These times roughly bracket the Christmas-New Year holiday period when the UFOs were seen. No other quakes in New Zealand were recorded between these two times.

These seismic tremors do not conclusively demonstrate that seismic activity is directly connected with the UFO reports, around the time of the Kaikoura UFOs, or that there was necessarily natural gas emission to fuel the UFOs. However, it does point in that direction and the possibility that there may have been gas emission preceding this seismic activity is a possible scenario.

It is worth noting that other UFOs were reported within this period. UFOs were seen on 16 December at Waikawa and during the periods of 20-21 December and 30-31 December. A release of natural gas emission during this time could have fuelled other vortices in other areas. On page 27 of his book he mentioned strange lights appearing in the atmosphere around the same time that puzzled several people. A farmer's wife saw a huge luminous UFO above some electric transmission lines just outside Blenheim.

The seismic data from the New Zealand Seismic Observatory is consistent with the notion of a so-called, "flap" of UFOs, which may be directly connected with seismic activity. I suggest that this seismic activity released a sufficient volume of fuel gas during this period to fuel UFOs seen in the "flap" (heightened activity) at the time. The build up of natural gas seems to take place before a quake as Gold has cited in association with luminous phenomena and earthquake activity. If Gold is right, natural gas detection seems to be another method to predict the occurrence of earthquakes.

Electrical and magnetic effects.

One remarkable aspect about the 1978/79 Kaikoura UFO was that it was detected by radar that required the existence of a pocket of localized ionization. This ionization was in the same location as visual sightings of the UFO. How can a vortex combustion theory, which is not an electrical theory, explain electrical ionization? It is known in flame studies that a flame can produce ionization (Gaydon and Wolfhard, 1953 and a later edition confirms this, (Gaydon and Wolfhard, 1970), where they stated that hydrocarbons generate a higher degree of ionization than some other combustible gases, like hydrogen.

In laboratory studies of flames charge separation has been known for some time. A thermal ionization reaction can split molecular species into a positive ions and negatively charged electrons. There are logical implications that follow from this charge separation. One prediction is that electric charge moving around in a swirling flame will generate a magnetic field. Electromagnetic radiation generated by thermal ionization should also be detected as radiation, in addition to thermal-generated electromagnetic radiation. The type of radiation emitted will depend on the frequency of oscillation of the charge in the flame. The ionization reactions which give rise to this thermal ionization effect will be dependent on temperature. The degree of ionization is likely to be variable and perhaps under high ionization conditions the fireball could be detected by radar detection. Conversely a very low ionization may not be detected by radar. This may explain why some fireballs are detected by radar and some are not. Another effect is that external electric fields, like that from a thunderstorm, can distort the combustion flame. The above electrical and magnetic effects should be detected in other fireball-type UFOs and ball lightning. This is yet another prediction based on the vortex fireball theory.

I now want to address the UFO observations by Frank McDonald of Waikawa. They are valuable from a scientific point of view because they appear to be a remarkably clear account, from a layperson's point of view, of the recirculation region of vortex-breakdown in a state of combustion.

Waikawa Bay UFO sighting

In the book, *The Kaikoura UFOs*, by Captain Startup and Neil Illingworth, the first chapter begins with a prominent UFO incident proceeding, by four days, the major Kaikoura UFO sighting. The event was on record as being seen by at least nine observers, like Frank MacDonald from Waikawa who managed to see the fireball through his binoculars. His detailed observations seem to show explicit evidence of combustion in vortex-breakdown.

Vortex-breakdown detected in the fireball?

The UFO settled itself above a ridge near Waikawa Bay. Frank described the shape as being like a *"sheath of wheat but with the stalks pointing upwards"*. This is rather like the "upside down pine tree" of the

Hessdalen lights. The sketch that was drawn by the witness of the UFO at the time of 12 55 am and was described as a "dome-shaped body" enclosing a diamond shape with heavy radiating beams emanating from it. The length of the object was estimated to be at least 400 feet in length. The light was an unusual bluish-green color but not like an electrical light. This key observation points towards a combustion flame. Frank MacDonald described the light in the following report:

"It was more or less like an open flame. It wasn't like electric light or a car's lights. It was flowing light like living light. The light was flowing and billowing outwards from the object. It never billowed inwards. It was billowing down towards the earth and it flowed out into a circle down at its feet and it was curling back. And these flames, these beams of light billowing out, they were practically the same color as the whole object."

I think the above description contains important observations particularly relevant to vortex fireball hypothesis and possibly even confirming it in this case. The open flame was like a "flowing light" would seem to suggest a combustion process. The billowing outwards and downwards and then curling back suggests a recirculation of air flow in the flame region consistent with the form of well established recirculation patterns seen in experimentally produced vortex-breakdowns. Frank MacDonald's description is a very rare sighting of a UFO, or indeed, of ball lightning, for that matter, where there is a good observational suggestion of a recirculation of air/fuel mix in vortex-breakdown. The reason why he could see such detail in the UFO was from the simple fact that he had binoculars.

The description of the flames from the Waikawa UFO is typical of many UFO sightings. Hervey (1975) reported on a Mrs Alcock from Federal, New South Wales who, while waiting at the Queanbeyan Railway Station, saw a rotating silver-colored UFO fly across the sky. It had blue flames (consistent with natural gas combustion?). As it moved in rapid circles flames were flung out, which looks very like the centrifuging action of a vortex.

The reference to rotation of the Waikawa UFO object is consistent with a vortex. The object was capable of "bouncing" up into the air to a 400 feet and back again to its original height of 100 feet above the ridge in about two seconds. This vertical rising and falling was repeated between 20 and 30 times. After the first "bounce" the glowing object apparently moved slowly, and drifted along the contours of the ridge in a northeasterly direction. The UFO would rise up over crests in the ridge, and fall in the troughs. What struck me about the movement of the UFO along the bush-clad ridge is that it parallels the behavior of naturally-occurring vortices which have a preference for following along the raised projections in the topography of the land. A vortex hovering over a promontory may explain cases where a fireball UFO is able to remain stationary, yet vertically above a ridge or hill crest.

The remarkable speed of vertical ascent and descent, could, in my opinion, only come from vortex-breakdown shifting vertically with changing flux of air currents entering the vortex. This rapid up and down movement of vortex-breakdown along a vertical axis is something I have seen in my vortex generator. The overall air flow pattern obviously did not vary too much, except perhaps, the air flow over the ridge and up into the vortex.

At one stage, the object tilted at such an angle, that Frank could see right up into what he described as a "machine".

"I could see a circle within a circle. Now the outside was a tunnel right around it. And there was another big circle in the middle of it (the dome he saw earlier when the machine was vertical). Whether the diamond shape was glassed in or not I don't know. The outer tunnel was connected to the dome by three smaller tunnels....And where these tunnels went into the outer tunnel they had big red cowlings over the end of them. These red sections, the diamond on the dome and the cowlings looked as though the machine could have been really hot. The rest of the machine was a silver-red color. Now I could see that the light from the bottom came up between these circles and around the outside of them. And it gave the top and all the tunnels a light pinkish sort of color."

Frank MacDonald reported on three subsidiary "tunnels" which are probably vortex funnels in a vortex splitting mode. Later the object was reported to distinctly change in shape and intensity. The upside down wheat sheath transformed to a very bright sphere with a cap-like shape on the top. In my opinion the dome and the red cowlings were vortex-breakdown recirculation regions which were part of the vortex splitting formation as discussed earlier. The shape change from an inverted wheat sheath to a globe is capable of being explained by vortex-breakdown shape changes.

Another Argosy UFO sighting

The Christchurch Press reported on Thursday the 15 March, 1990 of a UFO that was seen, at around 3 am, from an Argosy aircraft, en route from Auckland to Christchurch. The Argosy crew had been flying at 13 000 ft, when a light aircraft coming out of Palmerston, contacted the plane and reported a strange light. At first the Argosy crew could not see it, but not long after, it was then sighted. The pilot Grant Jolley described the light as having an apparent brightness somewhere between six and seven times that of stars in the night sky.

The light changed colors from white to red and green. The position of the object was between the two aircraft and below the tops of the Rimutakas. Mr Jolley said that the light was definitely not Venus. The object was observed to move on a course parallel to both aircraft, as in the case of the Kaikoura UFO.

THE 1988 NULLABOR PLAINS UFO

The Nullabor Plains UFO event made world headlines and had a profound effect on an ordinary Australian family traveling in a car. They had no idea what they encountered. From my reading of the Jenkin (1988) article in the *New Zealand Listener*, I found several observations in agreement with the ball lightning vortex burner hypothesis.

The story of the event goes like this. Faye Knowles and her three sons were driving across the Nullabor at about 3.30 in the morning on the 20th January, 1988. There was no moonlight on the night of the sighting. The main vegetation on the Nullabor (literally "no tree") is mallee scrub which is a kind of eucalyptus plant.

As the Knowles family car was motoring along the outback road they suddenly encountered an object that was jumping around and "playing" with them. Faye then described how the UFO suddenly transformed into a large yellow sphere. Patrick, one of her sons, said that as they approached the UFO it bolted upwards and seemingly vanished. But not for long because as the car turned around to escape the UFO, it began to follow them. This could be an example of the "countermotion" behavior I talked about earlier with the vortex-breakdown responding to perturbations of the air flow by moving objects in the vicinity of the vortex. Apparently when they tried turning to get away from the object, it always appeared as a "tear drop" shape in front of them. After some time they began driving back to Perth and they thought that the thing was gone. But suddenly to their horror, the object was on top of the car and making an unusual vibrating noise.

The vibrating noise is consistent with reports by those who have had experience with tornadoes. All the mechanical effects point to a mini tornado or dust devil lifting the car off the road. The air would flow along the ground boundary layer up into the vortex core and into the recirculation region of the fireball. Since the fireball was above the car then the main reverse flow would also be higher up. So the flow would be upward to lift the car from the underside and then flow around the cars body to move up to the fireball. The lift force must at least balance the gravitational force on the car. A "ball park" estimate of the air pressure from the high speed air on the underside (eg area = 8m²) of a one Tonne (1000 kg) car with four passengers each weighing 75 kg can be made. Using P=F/A (where P, F, and A are pressure, gravitational force and area respectively) yields a pressure of about 1625 Nm-².

Fortunately for the occupants of the car the fireball was above the car. Had the ball descended it could have had serious consequences to members of the Knowles family. In fact it probably was not a good idea to open the window. According to the vortex-breakdown theory this would mean a sudden reduction in the air flow feeding into the fireball above. Experiments on vortex-breakdown would suggest that the fireball would have a tendency to shift down because of this type of air flow perturbation. The descending fireball could well have burnt any combustible objects in its path. Hence this encounter was extremely dangerous given the size of the vortex burner. There was nothing benign about the event and it must have been a terrifying ordeal for the family. Luckily the family escaped harm perhaps because the car effectively acted as a projection from the ground, rather like a ridge, so that the lower part of the vortex air could flow up, around and under the car while the breakdown zone was above. Still the breakdown zone is sensitive to slight changes in air flow. Maybe at these greater volumetric air flow rates small disturbances might not show as dramatic a shift in the axial position of the breakdown zone, as in much smaller vortices in vortex chambers. Combustion may also have helped to stabilize the axial position of the fireball in this large desert vortex.

When Patrick opened his window there was a sudden pressure release. The car interior sucked in soot and smoke-laden air through the open window was most likely from air flow into the car when the window opened. The inside of the car was brightly illuminated by the strange light above. This light was likened to a laser beam. The physical sense of a pressure change and the accompanying air flow is explained by a fast moving air of the low pressure vortex moving up and around the curved body of the car providing a force on the window in a fluid momentum transfer effect described in fluid dynamics. When someone opens the window this force would be felt and then the air flow would change direction and move sideways into the car and out again to move upwards.

All sorts of explanations were put forward for this Nullabor UFO-from street lamps to car lights, and "exotic lightning". A carbonaceous meteor was proposed by physicist, Glen Moore of Woollongong University. But these theories have difficult problems in fully accounting for all the observed effects, as reported by members of the Knowles family. One difficulty with the carbonaceous meteor theory is how to explain the car lifting off the ground because real meteors are apparently incapable of doing this.

Furthermore the electrical lightning theory has problems. What about the very puzzling observation that the car had apparently no electrical burn marks on the paintwork? Graham Henley commented on this too. He questioned why the car's paintwork was not burnt. If it was an electrical phenomenon why was there no evidence of electrical effects experienced by the victims, such as shocks, or any other evidence of electrical burning? There was a bakelite smell which is suggestive of electrical burning. Alternatively it could have been combustion exhaust gases from heated minerals associated with the silica sand. Pieces of organic material, such as small twigs heated or burnt would end up as carbon would explain the soot observation. Indirect evidence for a combustion explanation comes from the observation of Faye Knowles when she put her hand out the window and she touched a black "suction cup" and it burnt her. This burning sensation was not accompanied by an electric shock that certainly poses a problem for an electrical theory.

The black "suction cup" description is quite puzzling but I suspect it has something to do with the air pressure effect from the lower part of the vortex lifting the car up. In effect, there was a physical equilibrium in which the weight of the car, and its passengers and cargo balanced the lifting force of the vortical air flow while the combusting recirculation region of the fireball lay further up. This explanation would also explain the absence of heating effects to the paintwork of the car, but I suspect there was a fair amount of radiative heat transfer but insufficient to burn the paint. As I have said it was fortunate *that* the burning part of the vortex was sufficiently above the car in the vortex-breakdown region to prevent excessive heat transfer to the car and its occupants. A minimum lifetime of this fireball object can be estimated from the *Listener* report. The object was first seen around 3.30 a.m. The object was still seen up until day break which have been around 5.15 a.m., as this was time was mentioned in the report. The lifetime of the object would then have to be at least 1 hour 45 minutes. The source of the fuel gas would have to last at least this length of time. It is highly likely that the object lasted longer than this because the Knowle's family drove off around daybreak with the light still in full view. I recall a staff member of the Canterbury University Physics and Astronomy Department who suggested that the Nullabor Plains UFO was merely a hoax. He said the UFO was probably a helicopter which stirred up dust and lifted the car off the ground. Even the observation of the strange circular light still visible after daybreak would most certainly rule out the helicopter theory. I believe that my vortex hypothesis is a reasonable and practical explanation of this UFO sighting and is consistent with the observational details as revealed by members of the Knowles family. This evidence includes the; rush of air from a pressure difference, the lifting of the car and the circumstantial evidence of burning such as smoke and soot. There was some suggestion of shape change because Faye said that after the thing finished jumping around it formed a yellow ball.

The vortex-breakdown theory is capable of explaining shape

changes. The metallic color seen near sun rise, compared with the nocturnal yellow color, may be due to a difference between nocturnal and day time viewing. In the latter situation the large amount of day light would tend to dominate the color seen during the night. The yellow color may be related to quantum optical emissions when the silica sand and carbon is heated rather than black body radiation.

Mt Gambier UFOs

The Knowles family UFO encounter is by no means a unique experience in the Nullabor Plains. In the Jenkins report the truck driver Henley is quoted as saying that UFOs have been observed in this area for years. Similar reports of UFO objects in this South Australian region reveal that these too might be explained by vortex fireball theory. I uncovered a much earlier report of a UFO in the same general area by *Cade* and Davis (1967). On page 108 of their book they cited a news release in the NZ Herald on May 22, 1963. Apparently a youth was chased by a mysterious object at around 9.30 pm, while driving his car between Glencoe and Mt Gambier near the Victorian border on May the 20th. There are definite similarities to this UFO light and the one seen by the Knowles family, especially in the way it moved above the car. As the youth moved toward the UFO (he was only around 20 meters away) it took off across the road. When he accelerated the light went above his car. Even though he drove between 50 to 60 mph, the light somehow managed to stay with the car. The vortex then attached itself to a projection from the ground, which vortices are known to behave in this way.

Other UFO reports in this area have been reported to the Mt Gambier police. A car carrying several passengers was "chased" by a light for several kilometers. Such UFOs seen on the Nullabor, near the township of Mt Gambier (a volcano), could be fuelled by natural gas emitted from a fissure in this desert area, since there do exist volcanic features which could well indicate the occurrence of natural gas clouds. It is this gas which could be sucked into dust-devil vortices wandering the desert floor to give the startling impression of a classic UFO.

Spontaneous human combustion

Another controversial scientific anomaly called "spontaneous human combustion" is an unexplained combustion phenomenon where people are mysteriously burnt usually while they are typically sitting in front of a fireplace with a chimney. Many hypotheses have been postulated, including the idea that a human body can excrete methane gas which is then ignited by an electrostatic spark and the body is abruptly set on fire. In many eyewitness accounts spontaneous human combustion has usually taken place adjacent to fireplaces. To explain this fireball vortex could get into a building by dropping down the chimney. So the connection of spontaneous combustion with fireplaces is no coincidence.

To add further credibility to this explanation fireballs behaving like ball lightning have been connected with this phenomenon (Randles and Hough, 1992). Some of these spontaneous human combustion fireballs have reported properties that have inspired some investigators to propose ball lightning as a candidate theory. The heat output of a swirling natural flame of a fireball vortex coming down a chimney or through an open window could easily be envisaged as causing considerable burns to the body. A vortex fireball could move towards a person and the air flow system may "attach" itself to a human body, in the same way that vortices are know to stop and hover over projections from the ground. And after the burning effects there would be no trace of the vortex, although there could be residual traces of natural gas or methane in the air. It would be useful to determine whether there was some kind of gas leak from either a public gas supply or from a source from the surroundings at the time of the event. This would provide at least one crucial test of the hypothesis.

Conclusion

The fireball theory described herein not only provides a solution to ball lightning, but also a set of mysterious lights, as well as a variety of related scientific puzzles; like bead lightning, the burning effects of tornadoes, presters, the Andes effect, the Tunguska meteor. Both ball lightning and a type of UFO (the one causing all the publicity) are in effect, examples of naturally-occurring atmospheric "swirl burners" or vortex fireballs.

This theory suggests there is no ball lightning, as such, like that imagined as a plasma separating off from a lightning discharge. The riddle of "ball lightning" is solved if the phenomenon is now understood in terms of combustion and atmospheric vortices. Meteorologists will be interested since atmospheric vortices are a well known and studied phenomenon. It is somewhat surprising that meteorologists have not considered the possibility of combustion in a columnar vortex but there is no reason to rule this out and various observations do support this contention.

Of course other science fields would need to contribute to the development of the theory. The role of ordinary lightning in the ball lightning puzzle plays a minor role and is not required in every fireball event. Nevertheless the theory predicts that lightning can simultaneously create the vortex, ignite some fuel gas, in the vicinity, to produce the classical 30-centimeter diameter form of ball lightning. The volcano seems to be even more capable of providing the requirements of an ignition source, a vortex and a combustible gas.

In the absence of a suitable explanation from the scientific fraternity, the extra-terrestrial UFO hypothesis has often acted as the default theory. However scientists no longer need to ignore the UFO issue through want of a satisfactory explanation. They can now dispense with unsuitable official proclamations that the UFO in question was a meteor, Venus, weather balloon, or some kind of optical effect, like that caused by ice crystals in a moon-lit night. It is now possible to inquire whether a UFO, in question, might in fact be a vortex burner.

Many human societies all over the world have apparently being seeing this kind of aerial vortex fireball for many centuries. It is only natural that such a phenomenon would have various mythical and extra-terrestrial interpretations placed upon it. Early observers with their active imaginations would have constructed their own non-scientific explanations of this fearsome fireball. In retrospect, what these observers have been seeing is mainly the flame zone of combustion within vortex-breakdown that can explain many shapes from pear-shaped such as the Kaikoura UFO, spherical or a cylinder.

A list of reports of mysterious lights or UFOs which I have come across could be accounted for by the vortex fireball explanation have all been included in this book. These UFO sightings, I believe, can now be retrospectively identified and accounted for within the vortex fireball theory. My sampling of sightings represents only a small sample from the annals of UFO events. There are many more such cases that I have not referred to, that might, upon application of the theory, turn out to be vortex fireballs.

The vortex fireball hypothesis is capable of being supported, refuted or modified by reliable observations of ball lightning or a certain type of UFO in the field. I believe there is already photographic evidence and written eyewitness accounts which already confirms the existence of the combustion fireball. I have mentioned these cases mainly in relation to UFOs since very few authentic ball lightning photographs exist. These events will eventually convince other scientific investigators of the worth of the theory. I am thinking here of the photographs like that of Campbell of the Texas UFO and also the Kaikoura UFO.

Hall (1972) discussed the criteria of "good" scientific theories. He originally made his comments in reference to competing vortex-breakdown theories, but his comments could equally apply to the situation of ball lightning. His criterion is derived from the work of the famous science philosopher, Karl Popper, for the testing of explanatory theories (Popper, 1969). Fundamentally a "good" theory of science must correspond with the facts as they are known. Furthermore it is not sufficient that it should just have explanatory capability, it must also accommodate the unusual, as well as the main features of the problem. The theory should also have arisen from a fundamental unifying concept and be resistant to refutation.

The vortex combustion theory has the features of a defendable explanatory theory of these fireballs because it deals with several difficult observations that other theories fail to address. The vortex fireball theory has one powerful unifying idea to explain the wide range of physical and behavioral properties of unexplained fireballs.

Furthermore the theory predicts that these fireballs should be found anywhere where there is a coincidence of combustible gas, vortices and an ignition source. Hence such lights should be seen in the vicinity of earthquake fault zones and connected with volcanoes because they have been associated with combustible gas emission. They have been reported in these locations. .

To conclude: The vortex fireball burner theory is a unified theory better able to address the wide range of sightings of anomalous lights that have previously been classified separately as ball lightning, UFOs, fox fire, earthlights etc than existing theories. The hypothesis could be used as a working hypothesis for scientists for future studies since the theory may provide an elegant and correct framework. There is still work that needs to be done in experimentation, field work and in developing a much fuller mathematical model. For detailed models of these fireballs the use of computational fluid dynamics using numerical combustion would facilitate further understanding. This theory can be directly applied to other scientific anomalies, such as the UFO problem, the Andes effect, the Tunguska meteor, and others described in this book. Quite likely there are other related scientific anomalies reported in the scientific literature that are directly connected with the vortex fireball. Ball lightning is really a misnomer according to the theory and really represents the missing science of atmospheric vortices.

ACKNOWLEDGEMENTS

My appreciation and thanks to the ball lightning researchers, especially to Stanley Singer, whose scholarly book assisted me in my investigation. He also helped me by having my paper "*A combustion hypothesis to explain ball lightning*" presented by Dr. Karl Nickel from Germany to a gathering of scientists at the Fourth International Symposium on Ball lightning which was held in Kent in 1995.

I would like to thank Dr Von Biel, my first supervisor, and Dr Valda McCann for allowing the *Kugelblitz* (German for ball lighting) investigation to get off the ground. My thanks to the Electrical Engineering Department for the use of their high voltage lab during 1988-89, especially to head technician, Bob Young with his infectious enthusiasm and stories of min mins over in Australia. He also provided me with a willing technician, Dave Barron, who assisted with the high voltage discharge experiments using a tall bank of high voltage capacitors.

A special thanks to Dr John Abrahamson of the Chemical and Process Engineering Department, who was my master's thesis supervisor in 1989-90. The thesis "*Ball lightning*" was successfully completed under his guidance. He also supported the communication of a paper representing the department, under Professor Earl, to the 1997, 5th International Symposia on ball lightning held in Tsugawa, Japan. Thanks to all the Chemical and Process Engineering staff who helped out in various ways including Dave Brown, Neville Foot, Ian Murray, Ron Boyce, Bob Gordon, Warwick Earl, Tony Allen. To Trevor Berry for his assistance in the thesis photographic work, and to Glen Wilson of the Stores Office.

Thanks to Alexander Deev for giving me an understanding and appreciation of flame processes from his specialized field of metallurgy, especially furnaces.

My thanks to the staff of the Physical Sciences Library and the Engineering Library of Canterbury University. I would also like to acknowledge the help of several authors whose work I have mentioned or quoted in this book mainly for review purposes. This body of published material forms an invaluable collection of scientific field observations, not always widely appreciated. A special mention and thanks goes to Vicki Hyde for publishing news of my research in *The New Zealand Science Monthly*.

My thanks to Professor Cole of the Geology Department at the University of Canterbury for acting as an AGU sponsor. He also examined an abstract on the experimental production of vortex fireballs intended for publication in Eos.

Permission was obtained from the following people for either direct quotes or pictorial material used in this book. I wish to acknowledge and thank them all for their kind assistance.

Helen Slane -*New Scientist*
Professor Jennison University of Kent, England
Professor P.C.W Davies, author and physicist from Australia
The organizers of the 1997 Tsugawa Symposium on Ball Lightning for permission to publish the photograph of the delegates.
Professor Jayant Gore, Purdue University
Dr Stanley Singer, President of the ICBL, Los Angeles
Associate Professor Hideho Ofuruton, Tokyo Metropolitan College of Aeronautical Engineering.
Dr Andre Von Biel, Physics Department, University of Canterbury, New Zealand.
Dr Geert Dijhuis, Secretary of the International Committee on Ball Lightning.William Corliss, author of *The Source Book Project* which deals with scientific anomalies.
Dr Brugge, editor of *Weather,* A journal of the Royal Meteorological Society of the UK.
Jerry Conley editor of the *Missouri Conservator*
Suzanne J. Wilson, author of an article on spook lights seen at Hornet, Missouri, US.

MEDIA RELEASES

- **The Christchurch Press** published an article on page 4 on 1 July 1998 called *"Burning inside tornadoes cause of UFOs"*.
- **Television New Zealand** The UFO story was presented nationally on the 2nd July 6pm news by TVNZ.
- Interviews with the author were by local Christchurch New Zealand TV stations CTV and CHTV (Brian Tucker) on 8th July, 1998.
- **The Wellington Newspaper- The Dominion** Feb 1999 made the front page. The article *"UFOs-Great Balls of Fire"* was the first time that a color photograph of a burning vortex was published in the popular press.
- **BBC interview** 16 September 1999 Wellington, Plimmer Towers building.
- **New Scientist** article *'All fired Up-What's making a weird glow at the center of a twister'*, May 1999, p17.
- The theory was aired internationally on **Discovery Channel and National Geographic and Discovery channel** film on Ball Lightning in 2001.
- **Chronicle University of Canterbury** 1978 *Kaikoura UFO finally explained- ball Lightning?*
- **Scoop Website www.scoop.co.nz** 17 March 2004 entitled: *"New Zealander first to explain UFOs"*. *This article was seen on several websites.*
- **Coleman's Vortex Fireball Website www.geocities.com/peter13857**

REFERENCES

Abrahamson, J., and J. Dinniss, 2000., Ball lightning caused by oxidation of nano-particle networks from normal lightning strikes on soil, Nature, **403**, 519.

Abrahamson, J. 2002. Ball lightning from atmospheric discharges.
Via metal nanosphere oxidation: From soils, wood or
metals. Philosophical Transactions of the Royal Society,
A 360 (Jan. 15):61-88.

Abrahamson, J. , Bychov A.V., Bychov V. L., 2002. Recently reported sightings of ball lightning: Observations collected by correspondence and Russian and Ukrainian sightings Philosophical Transactions of the Royal

Society, A 360 (Jan. 15):11-35.

Altshuler, M.D., House, L.L., Hildner, E., 1970, Is ball lightning a nuclear phenomenon? Nature, **228**, 545,546.

Aleksandrov, V.Ya., Golubev, E.M., Podmoshenskii, I.V., 1982, Sov. Phys. Tech. Phys., **27** (10), 1221, 1221.

Ansley,G., 1988, Mysteries seen in a different light,
    Newsday column, The Christchurch Press, Sept., p2.

Arago, F.,1837, Sur le Tonnerre, Annuaire, Bureau des Longitudes.

Argyle, E., 1971, Nature, **230**,179.

Ashpole, E., 1995, The UFO Phenomena- A Scientific Look at
    The Evidence for Extraterrestrial Contacts, Headline Book
    Publishing, London, 210pp.

Ashby, D.E.T.F., Whitehead, C., 1971, Is ball lightning caused
    by antimatter meteorites? Nature, **230**, 179,180.

Bach, E.W., 1993, UFOs From Volcanoes, Tenafly: Heritage
    Publications.

Barry, J.D., 1967, Ball Lightning, J. Atmos. Terr. Phys., **29**,
    1095, 1101.

Barry, J.D., 1980, Ball Lightning and Bead Lightning, Plenum Press, N.Y.

Barry, J.D., Singer, S., 1988, The Science of Ball lightning (fireball), ed. Y. Ohtsuki, World Scientific, Singapore.

Beér, J.M., Chigier, N.A., 1972, Combustion Aerodynamics Applied Sci Publications, London 264pp.

Beesley, T., 1873, The New Bottle Whirlwind of Nov 30th 1873, Symon's Monthly Meteorological Magazine, 8:149.

Berlitz, C., 1989, Devil's Triangle, Grafton Books, 204pp

Bonacina, 1946, The Widecombe Calamity of 1638, Weather **1**,122,125.

Botley, 1966, C.M., Weather **21**, 318.

Brand, W., 1923, Der Kugelblitz, Henri Grand, Hamburg

Brookesmith, P., 1980, The Age of the UFO, Orbis, London, 205pp.

Brown, R., 1990, Tornado in North Dorset, 21 December 1989, Weather, 45, 240.

Buckminster, et al, 1990, The Structure and Stability of
Non-Adiabatic flame balls, Combustion Flame, **79,** 381,392.

Burbridge, P.W., Robertson, D.J., 1982, Lightning associated phenomena and related geometric measurements, Nature, **300,** 623-624.

Buttlar von, J., 1979, The UFO Phenomenon, Sidgwick and Jackson, London, 204.

Cade, C.M., Davis, D., 1967, The Taming of the
Thunderbolts-The Science and Superstition of Ball Lightning,
Cawood, W., Patterson, H., 1931, Nature, Lond. **128,150.**

Chagger, T., 1982, Weather, **37**, No.3, 90, 91.

Campbell, S., 1982, Ball lightning at Crail-1968, Weather, **37**, No.3, 75,78.

Charman, W.N., 1971, After-images and ball lightning, Nature, **230**,576.

Charman, W.N., 1979, Phys.Rep., **54**, pp 26.

Charman, W.N. 1982, Ball Lightning: The unsolved problem, Weather, **37**, No.3, 66, 74.

Clarke, D.W., 1988, Spook lights, in Phenomenon-Forty Years of Flying Saucers, edited by J. Spenser and H. Evans, 308,315.

Coleman, P.F., 1988a, Journal 1 notes, MSc investigation.

Coleman, P.F., 1988b, Kugelblitz, MSc Physics Report, University Canterbury, pp25.

Coleman, P.F., 1990, Ball lightning, MSc Thesis, University of Canterbury, pp 224.

Coleman, P.F., 1993a, An explanation for ball lightning?, Weather, **48**, No.1, 30.

Coleman, P.F., 1995a, A combustion hypothesis to explain ball lightning, oral paper presented to 4th International Symposium on ball lightning.

Coleman, P.F., 1997, Vortex-breakdown burner hypothesis of ball lightning, Proceedings of the 5th International Symposium on ball lightning, 26-29 August, Tsugawa, Japan.

Coleman, P.F., 1998, Identification of the Tunguska "meteor", abstract in the Proceeding of the International Scientific Conference on Tunguska, Krasnoyarsk, June 30, 2 July, 1998.

Coleman, P.F., Abrahamson, J., 1999 Combustion Flame in a Tornado-like Vortex in a State of Vortex-breakdown Eos, Transactions American Geological Union, Spring Meeting.

Coleman, P.F., 2005, A unified theory of ball lightning and unexplained atmospheric lights, paper submitted to J. Sci. Exp.

Corliss, W.L.,1982, Lightning, Auroras, Nocturnal Lights And Related Phenomena-A Catalog of Geophysical Anomalies, The Source Book Project, PO Box 107, Glen Arm, MD 21057.

Crill, P.M., Barlett, K.B., Harriss, R.C., Gorham, E., Verry, E.S. Sebacher, D.I., Madler, L., Sanner, H., 1997, Methane flux for Minnesota peat lands, Global Biogeochemical Cycles (in Press).

Crozier, W.D., 1964, J. Geophys. Res., No. 24, 5427, 5429.

Crozier, W.D., 1970, J. Geophys.Res., 75, No. 24, 4583.

Davies, P.C.W., 1971, Ball lightning or spots before the eyes?, Nature, **230,** 576,577.

Davies, P., 1987, Great balls of fire, New Scientist, 24/31 December, 64, 67.

Deev, A., 1998, Personal communication.

Deng, Q., Jiang, P., Jones, L.M., Molnar, P. 1982, A Review (D.W.Simpson and P.G.Richards, eds),pp543-565.American Geophysical Union, Washington, D.C.

Devereaux, P., 1988, Earthlights,in Phenomenon -Forty Years of Flying Saucers, edited by J. Spenser and H. Evans, 316,328.

Dijkhuis, G.C., 1995, Report on the Fourth International Symposium on Ball Lightning, University of Kent at Canterbury, UK,25-27 July1995, Chairman,Prof.R.C.Jennison, The International Committee On Ball Lightning Newsletter 1,1996.

Docobo, J. A.; Spalding, R. E.; Ceplecha, Z.; Diaz-Fierros, F.; Tamazian, V.; Onda, Y. 1998, Meteoritics and Planetary Science, **33**, 57,64,

Egawa, M., 1997, Unpublished eyewitness accounts of kitsune-bi (fox fire) in the Tsugawa area, Japan.

Emmeneger, R., 1974, UFOs Past, Present and Future, Ballantine Books, N.Y. pp 207.

Endean,V.G.,1976, Ball lightning as electromagnetic energy, Nature, **263**,753,754.

Endean, V.G., 1993, Spinning electric dipole model of ball lightning, IEEE Proc. A, **140**, No.6, 474,478.

Eriksson, A.J., 1977, Video tape recording of a possible ball lightning event, Nature, **268**, 35-36.

Ezekial, The New Oxford Annotated Bible, RSV version,

Faler, J.H., Leibovitch, S., 1978, An experimental map of the internal structure of a vortex-breakdown, J.Fluid Mech.,**86**, part 2,313,335.

Flammarion, Thunder And Lightning, Chatto and Windus, 1905.

Franklin, B., 1753, Philosophical Transactions, Vol. **47**, 565.

Freier, G.D., 1960, J. Geophy.Res.,**65**, No. 10, 3504.

Gaydon, A.G.,Wolfhard, H.G.,1953, Flames and their Structure, Radiation and Temperature, Chapman and Hall, London, First Edition, 340pp.

Gaydon, A.G.,Wolfhard, H.G.,1970, Flames and their Structure, Radiation and Temperature, Chapman and Hall, London, Third Edition ,401pp.

Gaydon, A.G., Wolfhard,H.G, 1949, Spectroscopic studies of low pressure flames, in The Third International Symposium on Combustion Flame, and Explosion Phenomenon,504,505.

Galli,I., 1911, Raccolta e classificazione di fenomeni luminosi osservati nei terremoti. Bol. Soc. Sismol.Ital.,**14**, 221,447.

Gibilisco, 1984, Violent Weather, Hurricanes, Tornadoes/Storms,Blue Ridge Summit,P.A.Tab books,260pp.

Gold, T., 1987, Power From The Earth-Gas Energy For the Future, J.M.Dent and Sons, London, 208pp.

Good,T., 1989, Above Top Secret-The World Wide Coverup, Grafton, London, 590pp.

Goodlet,B.L. 1937, Lightning, J.Inst. Elect. Engrs., **81**, 1, 56.

Gore,J.,1998, Personal communication from Purdue University, US.

Grigorjev, A.I., 1988, The Science of Ball Lightning (fireball),ed. Y. Ohtsuki, World Scientific, Singapore.

Grigorgev A.I., Dunaeva, T.N., 1992, Bibliographical Guide To Ball Lightning Literature over 1982-1992, ICBL Article Series Nr. 1995/1,95pp.

Gupta, A., Lilley, P.,G., Syred, N., 1984, Swirl Flows, Abacus Press, Kent.

Gurney, C.,G., 1973, Unidentified Flying Objects, Abelard-Schuman, London, Fifth Impression, pp 143.

Hall, M.G., 1972, Vortex-breakdown. Ann Rev Fluid Mech.4: 195, 218.

Handel, P.H., 1989, New Approach to ball lightning, in Science of BallLightning,Y.H., Ohtsuki,ed., World Scientific, Singapore.

Handel, P.H., 1997, Theory of stationary non-linear ball lightning   system of fireball and atmospheric maser, Proceedings of the 5th International Symposium on Ball Lightning-Aug 26 -29,Tsugawa, Japan.

Hare, A.T; 1889, Globular lightning, Letter to Nature, **40,** 415.

Hervey, M., 1975, UFOs Over The Southern Hemisphere, R. Hale, London, 250pp.

Hill, E.L., 1960, Ball lightning as a physical phenomenon, J. Geophys. Res., **65**, No.7, 1947, 1952.

Hough, P., Randles,J., 1994 The Complete Book of UFOs: An investigation into alien contacts and encounters, Piaktus, London, pp304.

Hynek, J.A., 1972, The UFO Experience (A Scientific Inquiry), Abelard-Schuman Ltd, London.

Idso, S.B., 1974, Am.Sci., **62**, 530,541.

Ingle,W.H., 1971, Lightning's effects, New Scientist, **52**, 185.

Jenkin, D., 1988, UFO anguish, NZ Listener, April 9, 24,27.

Jennison,R.C., 1969, Nature, **224**, 895.

Jennison, R.C., 1971, Ball lightning and after images, Nature, **230**,576.

Jennison, R.C.,1997, Ball lightning: A general critique, Proceedings 5th International Symposium on ball lightning, 26 Aug-29 Aug 1997, Tsugawa, Japan.

Jones, J.C., Combustion Science-Principles and Practice, Millenium Books, 306pp.

Judd, A., Davies, G., Wilson, J., Holmes, R., Baron, G Bryden,I.,1997, Contributions to atmospheric methane by natural seepage on the UK

Continental Shelf Marine Geology,
140,3-4, 427,455.
Justice, A. A., 1930, Seeing inside a tornado, Mon. Weather Review, 58, 205.

Kangieser, P.C., 1954, Monthly Weather Review, June,147,152.
Kapitza, P.L., 1955, The nature of ball lightning,
Dokl.Akad.Nauti. SSSR, **101**, 245,248.
Keanan, D.,1990, Puzzling object sighted, Christchurch Press,pg.9, September,3,1990.
Khoo, B.C., Yeo,K.S., Lim, D.F., He, X., 1995, Vortex-breakdown in unconfined vortical flow, Expt. Therm. Fluid. Sci. **14,** 2,131, 148.
Kikuchi, H., 1995, Ball lightning, Handbook of Atmospheric Electrodynamics,volume 1, edited by Hans Volland, CRC Press, Boca Raton, 167,187.
Kikuchi,H.,1991, Electromagnetohydrodynamic vortices and corn circles, From Environmental and Space Electromagnetics,Springer Verlag, Tokyo, 585, 595.
Klass, P.J., 1966, Plasma theory may explain many UFOs, Aviation Week and Space Technology, **85**.
Kundt, W., 2001
The Tunguska Catasptrophe: A forming Kimberlite,The Tunguska 2001 International Conference Abstracts from,
www.geocities.com/CapeCanaveral/Cockpit/3240/abstracts01.htm

Lavan, Z.,Fejer, A.A., 1965, Luminescence in supersonic swirling flows, J.Fluid Mech., **23**, part1,173,183.
Lewis, M.P., 1988, Unusual lightning events, Weather, 272, 273.
Leonov, S.B., Pankova, M.B.,Tomilov,Yu.M.,1997, Absorbing halo of hydrocarbonic plasma fireball, Proceedings of the 5th International Symposium on Ball Lightning, ISBL 97, 26-29, August 1997, Tsugawa Town, Niigata, Japan.
Liedtke, H., Spatschek,1984, Stability of Plasma   cavitons,Proc.
Int. Conf.on Plasma Physics and Fusion, Lausanne, Switzerland.
Long,G., 1991, The Yakima UFO microcosm,Center For UFO Studies, 163pp.
Lowke, J.J., Uman, M.A., Liebermann, R.W., 1969, Toward a theory of ball lightning, J.Geophys.Res. **74**, No. 28, 6887, 6898.
Lowry, T.M., Cavell, A.C.,1968, Intermediate Chemistry, ninth   edition London, McMillan, 964pp.

Markson, R., Nelson, R., 1970, Mountain-Peak Potential-Gradient Measurements and the Andes glow, Weather, **25**, 350, 360.

Matsumoto, T.,1997, Ball lightning during underwater spark discharges and the Matsumae earthquakes, Proceedings of the 5th International Symposium on Ball lightning, Tsugawa, Japan.

Mathias, M.E., 1926, Les globes noirs et blancs sans lumiere propre, Comptes Rendes, **182**,32.

Maxworthy,T.,1982, The laboratory modelling of atmospheric vortices,Topics in Atmospheric and Oceanographic Sciences Intense Vortices,edited by Bengtsson/Lighthill,Springer-Verlag Berlin, Heidelberg.

McNally, J.R., 1966, Oak Ridge National Laboratory Report, ORNL-3938.

Meaden, G.T.,1989, The Circles Effect and its Mysteries, Published by Artetech Publishing Company, Bradford on Avon, 112pp.

Menzel, D.H., 1953, Flying Saucers, Putnam Co. Ltd., London.

Musya, 1931, Nature, **128**,155.

Nickel, K.L.E., 1989, A fluid dynamical model of ball lightning and bead lightning in, The Science of Ball lightning (fireball), ed. Y. Ohtsuki, World Scientific, Singapore.

Natsis, C., Potter, M., 1995, Reader's Digest Almanac of the Uncanny-Stories of the Supernatural Through the Centuries, Reader's Digest, Australia, pp46.

Ofuruton, H.,Ohtsuki,Y.H.,Kondo,N.,Kamogawa, Kato,M.,Takahashi,T., 1997, Nature of ball lightning in Japan, 5th International Symposium on ball lightning, 26-29 August, 1997, Tsugawa, Japan.

Ohtsuki, Y.H., 1989, The Science of Ball lightning fireball), ed. Y. Ohtsuki, World Scientific, Singapore.

Ol'khovatov, A. Yu.,1997, Explosions at Sasova, Russia, as examples of explosions of ball lightning-like formations endogenic origin, Proceedings 5th International Symposium of ball lightning, Tsugawa, Japan, 26 -29 Aug,1997.

Ol'khovatov, A. Yu.,1998, The tectonic interpretation of the 1908 Tunguska event (from http://www.tm.ru/tunguska/tunguska.htm

Pauley, R.T.,Snow, J.T., 1988, On the kinematics and dynamics of the 18 July 1986 Minneapolis tornado. Mon.Wea.Rev.,**116**,2731,2736.

Peckham, D.H., Atkinson, S.A.,1957, Preliminary results of low speed wind tunnel tests on a Gothic wing of aspect ratio1.0, Aeronaut.Res.Counc.,CP 508,16pp.

Pettigrew, J.D., 2003 The min min and the Fata Morgana- An optical account of a mysterious Australian phenomenon, Clin. Exp. Optom., 86,2,109,120.

Powell, J.R., Finklestein, D., 1970, Ball Lightning, American Scientist, **58**, 262, 279.

Randles, J. 1987, The UFO Conspiracy-The First Forty Years, Javelin Books, London,224pp.
Randles, J., 1992, UFOs And How To See Them, Anya Publications.
Randles, J.,Hough,P.,1992,Spontaneous Human Combustion, Robert Hale Limited, London, 224pp.
Rayle,W.D.,1966, Ball lightning characteristics, Natural Aeronautics and Space Administration, Washington, DC, Technical note, NASA TN D-3188, Jan. 1966.
Rutkowski, C.A., 1988, Geophysical alternatives, in Phenomenon -Forty Years of Flying Saucers edited by John Spenser, Hilary Evans, 301, 307.
Ryan, R.T., Vonnegut, B., 1970, Science, **168**,1349,1351.

Sarpakaya, T., 1995, Turbulent vortex-breakdown, AIAA7,10,2303.
Schofield, F.H., 1904, Remarkable Meteors, Monthly Weather Review, **32**, 115.
Schmidt,A., Mack, K., 1913, Das Süddeutesches Erdbeben vom 16, Nov. 1911, Würrtt, Jahrbücher f.Statist. u. Landeskde.,Jarhrg.1912, Heft I, 96-139.
Schmidt-Böcking H., Dorner R.,Jagutzki,Mergel,V.Spielburger,L.,Steibing, K.E.,Stohlker,T.,Schneider D., Schenkel,T., 1997, On the formation of "quasi -stable" hollow atoms towards a high power electric energy source, 133-136, Proceeding 5th International Symposium on ball lightning, Tsugawa, Japan
Silberg, P.A., 1962, J. Geophys. Res.,**67**,4941.
Silberg, P.A.,1966, Dehydration and burning produced by the tornado, J.Atmos.Sci., **23**, 202, 205.
Singer, S., 1963, The unsolved problem of ball lightning, Nature, **198**, No.4882, 745, 746.
Singer, S., 1971, The Nature of Ball lighting, Plenum Press, N.Y.
Singer, S., 1977, Ball lightning in the book Lightning by R.H.Golde, London Academic Press, 409,436.
Singer, S., 1997, The First Decade of International Symposia on Ball lightning, Proceedings of the 5th International Symposium on Ball Lightning, ISBL 97, 26-29 August 1997 Tsugawa Town, Niigata, Japan.
Smirnov, B.M., 1987, The properties of ball lightning, Physical Reports (Review section of Physics Letters) 152, No. 4, 177-226.
Smirnov, B.M., 1990 The properties of fractal clusters, Physics Reports (Review Section of Physical Letters), 188, No.1.
Smirnov, B.M. 1994, Long-lived glowing phenomena in the atmosphere, Joint Uspekhi-Fizicheskikh Nauk and Turpion Ltd (English Edition of

Physics-Uspekhi, **37**, (5), 1,4.

Smith,W.D., 1979, Seismology and related research in New Zealand, Seismological Survey Bulletin (N.Z.), S252.

Smith,W.D., 1983, Seismology and related research in New Zealand Seismological Survey Bulletin (N.Z.), S276.

Snow, J.T., 1984, J.Atmos.Sci. **41**, 2477, 2491.

Sotiropoulos, F; Yiannis V., 2001, The three-dimensional structure of confined swirling flows with vortex breakdown, J.Fluid Mech 426,155-175.

Sotiropoulos, F, Webster, D.R., Lackey, T.C.; 2002 Experiments on Lagrangian transport in steady vortex-breakdown bubbles in a confined swirling flow, J. Fluid Mech., 466, 215-248.

Spenser, Evans, 1988, Phenomenon-Forty Years of Flying Saucers,Avon Books, N.Y.,413pp.

Stakhanov, I.P., 1979, The Physical Nature of Ball Lightning, (Atomizdat, Moscow, in Russian)

Startup, B., Illingworth, N., 1980,
The Kaikoura UFOs Hodder/Stoughton,Auckland, 209pp.

Stenhoff, M., 1999, Ball Lightning- An Unsolved Problem in Atmospheric Physics, Kluwer Stott, Academic Publishers,360pp.

Stott M., 1984, Aliens Over the Antipodes, Space Time Press, Sydney, 251pp.

Sweeny, J. Beamont, 1996, A challenge to all alien life forms. The Christchurch Press, pg 2, Weekend Section, April 13, 1996.

Tambling, R., 1967, Flying Saucers-Where do they come from?, Horwitz Publications, Sydney, 158pp

Thorarinsson, S., Vonnegut, B., 1964, Whirlwinds produced by the eruption of Surtsey Volcano, Bull. Am. Met Soc.,**45**,No.,8, 440, 444.

Tomlinson, C; 1889, On some effects of lightning,Letter to Nature,**40**, 366.

Toepler, M.,1900 Ann.Phys. IV **2** 560:Meteorol.**2**.17,543.

Tootell, B., 1985 All Four Engines Have Failed-The True and Triumphant story of Flight BA 009 and the 'Jakarta Incident', Andre Deutsch/Hutchinson, New Zealand.

Turner, D.J. 1998, Ball Lightning and other meteorological phenomena Physics Reports, 293, 1,60.

Vonnegut,B.,1960, Electrical theory of volcanoes, J. Geophys. Res. **65**, 203.

Vasilyev, N.V., 1996 The Tunguska Meteorite problem 1996 Workshop on Tunguska, Bolgna University, Italy.

Vaughan, O.H., Vonnegut, B., 1976, Luminous electrical phenomenon associated with nocturnal tornadoes in Huntsville, Ala., 3 April 1974, Bull.

Am. Met Soc., **57**, No.10,1220,1225.
Vonnegut, B., Weyer, J.R.,1966, Luminous Phenomena in Nocturnal Tornadoes, Science,**153**, 1213,1220
Vonnegut, B.,1991, Presters,Weather,**46**, 360,361.
Vu, B.T., Gouldin, F.C., 1982, Flow measurements in a model swirl combustor, J. A.I.A.A.,**20**, No.5, 642, 650.

Wallace, R.E., Teng, Ta-Liang, 1980, Prediction of the Sungpan-Pingwu Earthquakes, August,1976, Amer. Seismol. Soc. Bull.,70,1199,1223.
Watson, L., 1997, Search for oil to start in Canty, The Christchurch Star, 31 October, 1997.
Weiss, P., 2002 From Science News, Vol. 161, No. 6, Feb. 9, 2002, p. 87.
Wessel-Berg, T., 2003, A Proposed theory of the phenomenon of ball lightning, Physica Du Nonlinear Phenomena, **182**, 3-4, 223-253.
Wilke, J.C., 1785, K.Vet.Acad.Nye Hand., **6**, 290.
Williams, R., Packenham, R., 1872, An extraordinary phenomenon, English Mechanic, **16**, p363.
Wilson, S.,1997,Spooklights,
Http:/www.state.mo.us/conservation/conmag/199701/2.html.

Zhang J., Megaridis, C.M., 1998, Soot microstructure in steady and flickering laminar methane/air diffusion flames, *Combustion and Flame,***112**,4,Journal Combustion Institute.
Zou, Y., 1988, in The Science of Ball lightning (fireball), ed. Y. Ohtsuki, World Scientific, Singapore.

www.ingramcontent.com/pod-product-compliance
Lightning Source LLC
Chambersburg PA
CBHW031840170526
45157CB00001B/371